The Perils of International Capital

Can foreign capital empower dictatorships? In this book, Faisal Z. Ahmed offers a unified theory of the impact of three prominent types of international capital – foreign aid, migrant remittances, and foreign direct investment – on the survival of dictatorships. Existing scholarship that examines different types of international capital in isolation misestimates their effects. The book's unified theoretical approach clarifies the channels through which a strategically oriented government can leverage each type of capital flow to finance two important instruments of nondemocratic politics: repression and patronage. The book's methodological approach takes seriously questions of causal identification, exploiting plausibly exogenous variation in capital flows to more precisely estimate their effects. In doing so, Ahmed introduces creative ways to turn the observable world into a quasi-experimental laboratory. The book's theory, case studies, and cross-national statistical evidence demonstrate *how* international capital can foster authoritarian politics. These findings challenge many existing studies and contribute to several important literatures in economics and political science.

Faisal Z. Ahmed is an assistant professor of politics at Princeton University. His research focuses on political economy and international economics. Ahmed's work on the political economy of foreign capital has appeared in journals in economics and politics, including the *American Political Science Review* and *Review of Economics and Statistics*. He has held fellowships at Nuffield College and the Hoover Institute and served as an international and macroeconomist at the White House Council of Economic Advisers and the Federal Reserve Bank of Chicago.

The Perils of International Capital

FAISAL Z. AHMED
Princeton University

CAMBRIDGE
UNIVERSITY PRESS

University Printing House, Cambridge CB2 8BS, United Kingdom

One Liberty Plaza, 20th Floor, New York, NY 10006, USA

477 Williamstown Road, Port Melbourne, VIC 3207, Australia

314–321, 3rd Floor, Plot 3, Splendor Forum, Jasola District Centre, New Delhi – 110025, India

79 Anson Road, #06–04/06, Singapore 079906

Cambridge University Press is part of the University of Cambridge.

It furthers the University's mission by disseminating knowledge in the pursuit of education, learning, and research at the highest international levels of excellence.

www.cambridge.org
Information on this title: www.cambridge.org/9781108488655
DOI: 10.1017/9781108771146

© Faisal Z. Ahmed 2020

This publication is in copyright. Subject to statutory exception and to the provisions of relevant collective licensing agreements, no reproduction of any part may take place without the written permission of Cambridge University Press.

First published 2020

Printed in the United Kingdom by TJ International Ltd. Padstow Cornwall

A catalogue record for this publication is available from the British Library.

Library of Congress Cataloging-in-Publication Data
Names: Ahmed, Faisal Z., author.
Title: The perils of international capital / Faisal Z. Ahmed.
Description: Cambridge, UK ; New York, NY : Cambridge University Press, 2019. | Includes bibliographical references.
Identifiers: LCCN 2019020883| ISBN 9781108488655 (hardback) | ISBN 9781108726856 (paperback)
Subjects: LCSH: Economic assistance – Political aspects – Developing countries. | Investments, Foreign – Political aspects – Developing countries. | Economic development – Political aspects – Developing countries.
Classification: LCC HC60 .A45225 2019 | DDC 338.9109172/4–dc23
LC record available at https://lccn.loc.gov/2019020883

ISBN 978-1-108-48865-5 Hardback
ISBN 978-1-108-72685-6 Paperback

Cambridge University Press has no responsibility for the persistence or accuracy of URLs for external or third-party internet websites referred to in this publication and does not guarantee that any content on such websites is, or will remain, accurate or appropriate.

To my parents, Suraiya and Ziauddin Ahmed

Contents

List of Figures		*page* viii
List of Tables		ix
Acknowledgments		x
1	The Politics of International Capital	1
2	International Capital and Authoritarian Survival: A Descriptive Overview	24
3	Foreign Rents and Rule	42
4	Aiding Repression	76
5	Remittances and Autocratic Power	103
6	Foreign Direct Investment in Militarism	126
7	Conclusion	152
References		167
Index		181

Figures

1.1	Total inflows of foreign aid, migrant remittances, and FDI to developing countries	page 4
1.2	Cumulative inflows of international capital, 2000–2015	5
2.1	Total foreign aid, remittances, and FDI worldwide, 1970–2015	28
2.2	Leader tenure and international capital in strong autocracies	33
2.3	Leader tenure and international capital in weak autocracies	34
2.4	Leader tenure and international capital in democracies	35
2.5	Marginal effect of international capital on leader exit at varying levels of autocracy	39
2.6	Marginal effect of international capital on leader exit at varying levels of autocracy, with country and year fixed effects	40
3.1	Tax effort and political rights	47
3.2	Public sector compensation and democratic governance	54
3.3	Government welfare payments and democratic governance	57
3.4	Defense burden and democratic governance	58
3.5	Defense burden in autocracies and democracies, 1986–2015	58
3.6	Corruption and democratic governance	60
4.1	US bilateral economic aid, 1960–2009	80
4.2	US economic aid to Somalia, 1960–1990	85
4.3	Fragmentation in the US House of Representatives and average US bilateral aid disbursements	88
4.4	Annual probability of receiving US aid (P_i) and average US bilateral economic aid	89
4.5	Annual probability of receiving US aid (P_i) and annual variation (standard deviation) in US bilateral economic aid (by country)	90
5.1	Oil prices and remittances	108
5.2	Foreign rents and executive constraints in Egypt	123
5.3	Foreign rents and executive constraints in Jordan	124

Tables

2.1	Top recipients (as a percentage of GDP) of foreign aid, remittances, and FDI: Country average over the 2000–2015 period	29
2.2	Composition of international capital, 1970–2015	30
2.3	Political institutions and the tenure of leaders since 1960	32
2.4	Marginal effect of international capital and political institutions on leader exit	38
3.1	Political repression in autocracies and democracies	51
3.2	Channels	63
4.1	Political institutions and political repression in Somalia, 1960–2010	83
4.2	US aid harms political rights	94
4.3	Evaluating the channels	97
4.4	US aid fosters authoritarian stability	99
4.5	Political survival in US aid recipients	101
5.1	Remittances and distance to Mecca	109
5.2	Executive constraints and political survival	111
5.3	Pretreatment characteristics between Muslim and non-Muslim countries prior to oil price shock	113
5.4	Remittances strengthen authoritarian politics	114
5.5	Remittances and public finances	117
5.6	Remittances raise corruption	119
5.7	Discounting other channels	120
6.1	Are successful and failed oil explorations similar?	134
6.2	New oil discoveries increase military spending in nondemocracies	136
6.3	Oil discoveries do not spur security competition	139
6.4	Oil discoveries increase FDI and military spending	140
6.5	Discounting other channels	142
6.6	Oil discoveries are not politically destabilizing	144
7.1	The perils of international capital: Summary of findings	154

Acknowledgments

This book has evolved from a "three-paper" dissertation. Initially, I did not envision writing a book. However, I am glad that I have. This would not have been possible without the support from many mentors, fellow researchers, as well as my friends and family. I would like to take this opportunity to thank them.

I owe a special debt of gratitude to my mentors: Laura Alfaro, Kerwin Charles, William Howell, Jon Pevehouse, Duncan Snidal, and Eric Werker. Each person has generously offered his or her time and expertise to helping me become a better scholar. I am especially appreciative of my collaborations with Laura and Eric. Since my graduate school days, we have coauthored papers on a variety of topics in political economy and international economics. I have learned a great deal from these projects. My work with Eric inspired the empirical approach in two dissertation essays, which were ultimately published (Ahmed 2012, 2013) and also feature in this book.

This book represents an endeavor to think critically about how international capital can affect public finance in developing countries and its effect on "politics" in different institutional settings. While the book presents a new argument, it builds on some of my prior research. In particular, portions of three previously published articles feature in Chapters 4 and 5: "The perils of unearned foreign income: Aid, remittances, and government survival" (in *American Political Science Review* 2012), "Remittances deteriorate governance" (in *Review of Economics and Statistics* 2013), and "Does foreign aid harm political rights? Evidence from U.S. aid" (in *Quarterly Journal of Political Science* 2016).

I am grateful for financial support from the Niehaus Center for Globalization and Governance at Princeton University, Nuffield College, and the Hoover Institute at Stanford University. Support from the Niehaus Center and Nuffield allowed me to work on the published articles that feature in the book. During my time as a National Fellow at Hoover, the tranquil

Acknowledgments

environment there and ample supply of Philz's coffee proved to be a fantastic setting in which to think, read, and write.

I am deeply indebted to many people who have provided helpful comments at various stages of this book project, especially on prior published work. At Princeton, I am grateful to my colleagues for their encouragement and feedback. Constructive comments from Helen Milner, Amaney Jamal, Joanne Gowa, Christina Davis, James Vreeland, Andrew Moravscik, Robert Keohane, Kris Ramsay, Jake Shapiro, Ethan Kapstein, and Melissa Lee have been especially helpful. I have also benefited from insightful feedback from scholars at other institutions. They include Ida Bastiaens, Chris Berry, Allison Carnegie, Kerwin Charles, Raphael Cunha, Adam Dean, James Fenske, Ben Graham, Robin Harding, Andy Harris, William Howell, Noel Johnston, Adeel Malik, Kevin Morrison, Paul Niehaus, Brent Nieman, Ralph Ossa, Jong Hee Park, Kyung Park, Jon Pevehouse, Pablo Pinto, Alberto Simpser, David Singer, Duncan Snidal, Dustin Tingley, Eric Werker, Matthew Winters, and Joseph Wright.

In addition to all these people, numerous others asked helpful questions and offered constructive comments at various forums over the years. These venues include presentations at the meetings of the American Political Science Association, the International Political Economy Society, the Midwest Political Science Association, the Northeast Universities Development Consortium, the World Bank's Annual Conference on Development Economics and seminars at the Center for the Study of African Economies (Oxford University), UC Berkeley, UC Irvine, London School of Economics, Nuffield College, NYU, University of Chicago, University of Illinois at Champaign-Urbana, University of Maryland, Princeton University, and Yale University.

In the final stages of this book writing journey, I benefited from a workshop at Princeton in December 2017. The generous feedback from Eric Arias, Jeffry Frieden, Joanna Gowa, Quan Li, Charles Lipson, Melissa Lee, Alastair Smith, and Xander Slaski helped improve the manuscript in all facets. I thank the Niehaus Center and the Department of Politics at Princeton for the financial support to make the workshop possible. Suggestions from two anonymous reviewers further improved the final manuscript. I also thank Robert Dreesen and the team at Cambridge University Press for a smooth publication process.

Finally, I thank my family for their love and support throughout my education and career. I am deeply appreciative of the many sacrifices my parents have made for me and my younger sister, Shafinaz. Growing up, they encouraged us to pursue our interests and to be inquisitive of the world around us. This curiosity and my father's encouragement to be an "all-rounder" (one of his favorite sports analogies from cricket) inspired me to study math and economics, and now apply that training to studying political economy. I dedicate this book to my parents. My good friend Indranil Sen-Gupta carefully read and helped revise various drafts of the manuscript. The book has less jargon (I hope) because of him! Last but not least, I thank my wife Saba for her love, constant encouragement, and willingness to read (and edit) the final drafts of the manuscript. With her help, this year will see the publication of my first book and the birth of our first child.

I

The Politics of International Capital

Attracting financial capital is essential for economic growth in developing countries, but tragically can often foster nondemocratic politics. Consider, for example, the impact of foreign aid. Since 2008 Ethiopia has been one of the largest recipients of US aid in Africa, averaging around $80 million per year. While the aid is intended to foster economic development, practitioners are growing increasingly wary of its political ramifications.[1] Before the Ethiopian national election in 2010, foreign donors were charged with "subsidizing a regime that is rapidly becoming one of the most repressive and dictatorial on the continent." Western aid officials "seem reluctant to admit that there are two Prime Minister Meles Zenawis. One is a clubbable, charming African who gives moving speeches at Davos and other elite forums about fighting poverty and terrorism. The other is a dictator whose totalitarianism dates backs to Cold War days."[2]

Ethiopia is hardly the exception in terms of aid being "misused" by governments for nefarious political purposes. During the Cold War, foreign aid funded General Siad Barre's highly repressive dictatorship in Somalia. "Foreign aid," observed one Somali scholar, "provided the glue that held the system together in spite of internal waste and corruption."[3] The emergence of "new" donors after the Cold War has seemingly not changed the politically pernicious effects of aid.[4] One recent study, for example, finds that African leaders disproportionately channel Chinese aid to finance "white elephant"

---

[1] In *Development with Freedom*, Human Rights Watch, for example, provides extensive documentation about how the Ethiopian government uses aid to repress. Based on interviews with 200 people in 53 villages and cities throughout the country, the report concludes that the Ethiopian government uses aid as a political weapon to discriminate against nonparty members and punish dissenters, sending the population the draconian message that "survival depends on political loyalty to the state and the ruling party." For example, more than fifty farmers in three different regions said that village leaders withheld government-provided seeds and fertilizers (and even microloans) because the farmers did not belong to the ruling party; moreover, some were asked to renounce their views and join the ruling party in order to receive assistance.
[2] Epstein (2010).   [3] Adam (1999), 175.
[4] For example, see Bermeo (2017) on the landscape of aid after the Cold War.

projects, especially in the leaders' birthplaces.[5] This fits a broader narrative that aid is misused, leading many experts and practitioners to conclude that despite more than fifty years of development assistance, foreign aid has been ineffective and "dead."[6]

The misuse of foreign aid by recipient governments has led to a championing of other types of external development finance that seemingly bypass these governments altogether.[7] One such flow is migrant remittances, which flow to *households* and have been heralded as a "new development mantra."[8] With regard to their political ramifications, remittances – as a source of household income – have the potential to shape individual political behavior and may be a conduit for political liberalization.[9] This view has percolated to the upper echelons of public policy. In a speech promoting human rights in Cuba, for example, President Obama declared that "increasing the flow of remittances and information to the Cuban people" constitutes "measures that decrease the dependency of the Cuban people on the Castro regime" and that efforts to promote "contacts between Cuban-Americans and their relatives in Cuba are means to encourage positive change in Cuba."[10] For example, by decreasing the population's dependency on the regime (e.g., through its provision of lower quality public education), remittances can help free government resources to "reward" its supporters through state patronage. Yet, despite President Obama's positive disposition to remittances as a source of democratization (i.e., "positive change"), Cuba has successfully transitioned from the dictatorial rule of Fidel Castro to that of his brother, Raul. In this instance, an individual's detachment from the ruling authoritarian regime may have actually and tragically empowered it.

Of course, migrant remittances are not the only type of external capital that bypasses a government's coffers. Another is foreign direct investment (FDI), capital that accrues to *firms* in host countries. Unlike other types of foreign capital that can enhance the productivity and profits of firms (e.g., portfolio investment), FDI is unique in its longer-term investment horizon.[11] By definition, FDI entails an ownership stake in an enterprise abroad, frequently with investments in immobile factors of production, such as factories, oil wells, and the training of workers in the host country.

The "stickiness" of FDI can be a double-edged sword. On the one hand, in an effort to attract FDI, governments often undertake reforms that increase transparency, reduce the state's powers to expropriate property, and raise the government's commitment to the rule of law. On the other hand, the stickiness

---

[5] Dreher et al. (2016). [6] Easterly (2006); Moyo (2009).
[7] An exception is Dietrich (2016), who documents the recent trends of aid "bypassing" governments. Bypassed aid, however, remains a very small share of total foreign aid.
[8] Kapur (2004). [9] Lipset (1959); Acemoglu and Robinson (2006). [10] Obama (2009).
[11] Ahlquist (2006) provides an excellent account for the longer-term investment horizon of FDI and its political implications.

of FDI may advance a government's development goals and political strategies, especially in nondemocracies. For example, on coming to power in Brazil in 1965, the military actively courted FDI as a means of raising economic growth and gaining the support of the country's business elite.[12] The resulting "Triple Alliance" among the military, domestic business elites, and foreign investors sustained the military's dictatorial rule in Brazil for nearly two decades. Such a governing strategy is not unique to Brazil: dictators in Indonesia and Egypt have also successfully leveraged FDI to sustain their dictatorships.[13]

These anecdotes – foreign aid in Ethiopia, remittances in Cuba, and FDI in Brazil – suggest that foreign capital can prop up dictators. But how widespread is this phenomenon? What are the channels through which governments can harness these different "types" of capital flows to enhance their survival prospects? After all, these flows accrue to different actors in a receiving country (i.e., the government, migrant households, and firms), which can require different government strategies. And from an empirical perspective, given that a country's socioeconomic and political characteristics affect its inflows of foreign aid, remittances, and FDI – thus introducing issues of endogeneity – is it possible to gauge the *causal* impact of international capital on authoritarian politics? This book addresses these questions.

From the perspective of economic welfare, answering these questions is important because financial capital – which can also augment technological and human capital – is essential to economic growth and progress, especially in most capital-scarce developing countries. For these countries, foreign capital can help fill shortfalls in domestic saving (e.g., through foreign aid), technology (e.g., through FDI), and investments in human capital (e.g., through remittances).[14] These capital flows have grown tremendously for developing countries in the past forty years.

Figure 1.1 plots the aggregate amount of foreign aid, remittances, and FDI to 130 developing countries since 1970.[15] In the 1970s, developing countries received about $31.5, $11.5, and $11 billion in foreign aid, remittances, and

---

[12] As described by Evans (1979), foreigners tended to invest in firms and sectors controlled by large and powerful business leaders/families in Brazil.
[13] For Indonesia, see Robison (1986). For Egypt, see Marshall and Stacher (2012).
[14] Among development economists, a "gap" in investment due to insufficient national savings is a leading reason why developing countries are poor. Foreign aid is a means to fill this savings/investment gap (Chenery and Strout 1966). In contrast, FDI injects financial capital to firms and is frequently accompanied by the transfer of new technologies and firm-specific assets (e.g., managerial expertise, patents) that enhances the productivity and profits of domestic firms in receiving countries. Remittance income can be spent by migrant households in numerous ways, including on investments in education and/or better health care, both of which enhance an individual's human capital (Acosta et al. 2007).
[15] Appendix Table A1 lists the sample of developing countries that receive nonnegative amounts of aid, remittances, and FDI. These countries span Central and South America; the Caribbean; Africa; the Middle East; Central Europe; and Central, South, and East Asia.

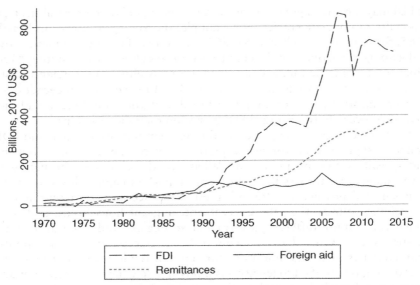

FIGURE 1.1 Total inflows of foreign aid, migrant remittances, and FDI to developing countries.
Note: Total inflows of foreign aid, remittances, and FDI to developing countries are based on the author's calculations. Data from World Development Indicators. See Table A1 for the sample of countries.

FDI, respectively.[16] All three capital inflows grew modestly in the 1980s, with especially dramatic increases in remittance and FDI since the 1990s. By 2010, aggregate inflows of aid, remittances, and FDI to developing countries amounted to $84, $312, and $711 billion, respectively. Since 2010, FDI inflows have exceeded aid and remittances, but tend to be more volatile. In contrast, foreign aid and remittances are sometimes more stable forms of external finance.

Foreign capital varies across countries. Figure 1.2 shows the average "intensity" of international capital for several developing countries. For each country in the figure, the corresponding numerical value is the average amount of aid, remittances, and FDI (as a share of GDP) a country received between 2000 and 2015. For a few countries, international capital comprises a significant share of their national economies. For example, foreign capital comprises about 40 percent of Afghanistan and Tajikistan's national income. At the extreme, only 7 percent of Liberia's income is produced domestically. Although Liberia is an outlier, international capital is nonetheless an important source of national income in many other countries. For instance, about one out of every five dollars of national income in Honduras, Jamaica,

---

[16] All figures are in 2010 US dollars.

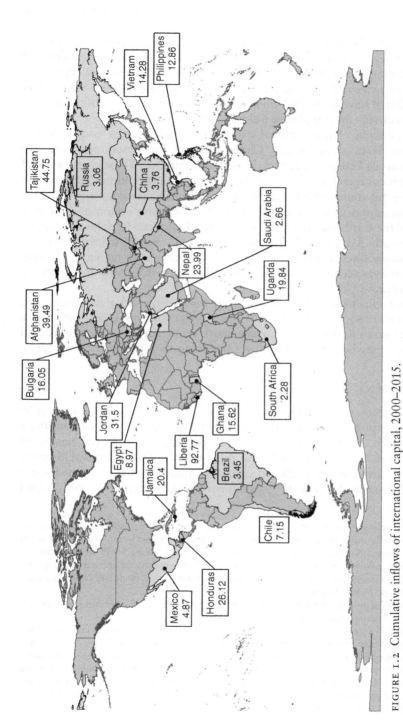

FIGURE 1.2 Cumulative inflows of international capital, 2000–2015.
Notes: The figure depicts the cumulative inflows of foreign aid, remittances, and FDI as a share of gross domestic product (GDP) for an arbitrary selection of developing countries across regions of the world (South America, Africa, the Middle East, and Asia). Each number reports the average share between 2000 and 2015. Calculations use data from the World Development Indicators.

Nepal, and Uganda comes from overseas. For "larger" developing country economies, international capital unsurprisingly comprises a smaller share of economic income (e.g., Brazil, China, Russia). These low shares, however, mask each country's extremely large aggregate (nominal) capital inflows. Since 2000, China and Brazil have been the beneficiaries of more than $2.7 trillion and $1 trillion, respectively, in cumulative inflows of foreign capital.[17]

The *composition* of international capital also varies across countries. For example, East Asian countries (e.g., China, Vietnam, Indonesia) tend to receive higher amounts of FDI compared to aid and remittances.[18] These countries tend to be middle-income countries. In contrast, foreign aid and remittances tend to be a more important source of external finance in lower-income countries. Migrant remittances are important sources of foreign capital in many countries in Central America and the Caribbean (e.g., Honduras, the Dominican Republic, Jamaica), as well in South Asia (e.g., Nepal, Bangladesh). In many African countries, foreign aid constitutes a significant share of each country's aggregate income. Since 2000, aid has accounted for more than 10 percent of Somalia and Uganda's national income.

Together, Figures 1.1 and 1.2 demonstrate that foreign aid, remittances, and FDI are important sources of capital for developing countries. I will now discuss how these different flows – aid to governments, remittances to households, and FDI to firms – may be exploited by governments to their political advantage.

1.1 ARGUMENT

This book argues that international capital can foster authoritarian rule: international capital offers opportunities to strengthen autocratic *institutions* and *governance*, with the aim of helping prolong the political *tenure* of dictators.[19] Building on revenue-based theories of political survival, I argue that autocrats can use international capital to finance strategies that bolster their political survival.[20] For all governments, these strategies broadly entail

---

[17] Inflows are measured in constant 2010 US dollars. Per annum, China and Brazil have received around $182 billion and $69 billion respectively since 2000. India, with annual inflows of about $75 billion since 2000 (equal to 5 percent of GDP per annum, or $1.1 trillion in cumulative foreign capital inflows), represents another recipient of large amounts of international capital.

[18] The Philippines is an exception, as it receives significant amounts of remitted income from the Persian Gulf and North America.

[19] Thus, measures of political institutions, governance, and political survival are important "dependent variables" investigated in this project.

[20] If political actors are rational and "optimizing" their decision-making in equilibrium (a common assumption in positive political economy), then additional inflows of capital should *not* threaten the political survival of dictators (as argued in this book). Empirically, this implies that additional amounts of international capital should *not increase* the dictator's probability of losing political office. This is equivalent to observing either a statistically significant decrease in the likelihood of losing office *or* no discernible impact from an additional amount of international capital received in autocracies.

1.1 Argument

funding political "repression" and "loyalty," which in turn can affect the quality of governance (e.g., lead to corruption) and features of political institutions (e.g., constraints on executive authority). However, the importance of each strategy is greatly influenced by the country's underlying quality of democratic political institutions. In particular, relative to democrats, autocrats will allocate a greater share of their revenues to financing repression and accumulating loyalty.

Whereas foreign aid is given from one government to another, the transfer of funds is different for remittances and FDI. A government can siphon foreign aid directly into its revenue base, constituting an *income effect*. On the other hand, largely untaxed remittance income – which accrues to households – can allow a government to diminish its provision of welfare goods (e.g., health care, pensions) and divert those expenditures to policies that fund repression and accumulate loyalty, representing a *substitution effect*. By contrast, FDI can increase a government's revenue in two ways – by creating opportunities for government rent extraction (*rentier effect*), and through higher wages that a government can tax. Moreover, the institutional features in nondemocracies, which incentivize autocrats to finance repression and patronage, magnify the utility of implementing these strategies in countries with authoritarian institutions.

As a consequence, this generates a *conditional theory* of international capital and government finances across regime types: relative to democrats, international capital can allow autocrats to fund various strategies that strengthen their prospects for political survival. As the empirical chapters demonstrate, some of these strategies include the deterioration of political rights (funded by foreign aid), the diminution of an autocrat's constraints on power (induced by remittance income), and an augmentation in military spending (associated with inward FDI).[21] Thus, international capital can affect both the quality of *governance* (e.g., corruption, political rights) and political *institutions* (e.g., executive constraints) in dictatorships, both of which affect a leader's political survival.

The unified approach of this project reveals a broader detrimental impact of international capital than the study of any one capital flow, as developing countries are the recipients of varying amounts and compositions of foreign aid, remittances, and FDI. Scholars of foreign aid rightfully debate whether autocrats siphon aid directly to pay off loyal supporters, but this debate ignores

---

[21] The ability of dictators to harness capital *inflows* to their political benefit parallels arguments that *outflows* can constrain dictators and foster democratic reforms (Boix 2003; Freeman and Quinn 2012). For instance, Freeman and Quinn argue that mobile capital can generate pro-democratic effects in autocracies with greater capital account openness and inequality. However, the greater "stickiness" (i.e., lower prospect of "exit") of foreign aid, remittances, and particularly FDI weakens the potentially countervailing effects of mobile capital. Chapter 3 discusses other potential countervailing effects and the strategies employed in this book to address these concerns.

the substitution effect that remittances can provide to autocrats, and also ignores the ways in which FDI can generate rents for autocrats. A unified approach is theoretically and methodologically important because countries receive varying levels and compositions of these capital flows.[22]

1.2 EXISTING RESEARCH

Understanding how distinct types of international capital can affect authoritarian politics in a unified framework is largely absent in the extant literature. Rather, scholars tend to theorize on and analyze the political consequences associated with each capital flow in isolation.

Foreign Aid

While a large literature exists on the political economy of foreign aid, two strands are particularly pertinent for evaluating the effect of foreign aid on political survival and institutions.[23] The first examines the conditions under which aid can be "effective" in fostering economic development. The workhorse models posit that aid can raise economic growth by filling the "investment gap" in recipient countries. An extensive empirical literature examines whether "aid causes growth" but a consensus on this issue remains elusive.[24] However, as these studies have progressed over time, scholars have increasingly recognized that democratic governance in recipient countries can mediate the effect of aid on growth. In particular, aid is more likely to engender positive development outcomes in democratic settings.[25] Democratic governments have stronger incentives to spend aid on public goods. In many instances, democratic governments are also more likely to undertake economic and political reforms "tied" to aid disbursements. Consequently, the effect of aid on beneficial economic outcomes may be "conditional" on the recipient's underlying polity type.

Of course, the relevance of politics in the aid–growth nexus may be more direct. The second strand posits that aid can directly impact political survival and institutions; these, in turn, can affect a recipient's economy. For instance,

---

[22] From a methodological perspective, many studies looking at international capital flows in isolation suffer from omitted variable bias. For instance, if poorer countries tend to receive higher amounts of aid *and* remittances, omitting remittances from an econometric model that regresses foreign aid on political survival may overstate ("upward bias") the effect of foreign aid.

[23] For example, many studies examine the rationale and determinants of aid disbursements (e.g., Morgenthau 1962; Alesina and Dollar 2000; Werker 2012), while others gauge the impact of aid on "politics" (e.g., Bueno de Mesquita and Smith 2010a; Bermeo 2011; Ahmed and Werker 2015).

[24] Roodman (2007) analyzes several studies to highlight the "fragility" in empirically evaluating the effect of aid on growth.

[25] Burnside and Dollar (2000).

1.2 Existing Research

Morrison (2015) posits that foreign aid represents a source of nontax revenue that a government can spend on policies that lower the prospect of government termination and ward off institutional change.[26] Moreover, foreign aid can diminish a government's tax effort, and concomitantly, its willingness to be politically accountable to its population.[27] Viewed in this light, foreign aid as a source of nontax revenue is akin to government revenue generated from the exploitation of natural resources.[28] In a panel of 108 countries from 1960 to 1999, Djankov et al. (2008) found that foreign aid deteriorates democratic institutions and that the "political curse" associated with aid may be larger than that related to oil.

Some scholars, however, remain skeptical regarding whether aid is *like* oil. One such area of differentiation is a donor's "intent."[29] For instance, during the Cold War, donors often disbursed aid to "buy" allegiance from their recipients while ignoring the political ramifications of that aid. After the Cold War, the strategic interests of Western donors changed: they now increasingly "tie" aid disbursements to democracy-enhancing projects and reforms in recipient countries.[30] A donor's preferences are not necessarily at odds with the "aid as nontax revenue" argument. For example, foreign aid from donors that have a pro-democracy intent could help a government finance its repressive capacity while still being tied to political reforms in other areas. Assessing the net effect of aid on democratic governance therefore remains an important empirical question, but one that is fraught with challenges, such as endogeneity. Existing studies frequently fail to mitigate concerns with endogeneity bias and do not provide robust causal effects of the aid on democracy.[31] In this book, I tackle this challenge head on.

Migrant Remittances

Scholarly skepticism about the impact of foreign aid on governance has led to an academic championing of workers' remittances as a means to improve democratic participation. For many scholars, remittances constitute money that does not directly accrue to a government and is deemed "value neutral." As such, conventional scholarly wisdom portends that remittance income can improve the quality of democracy and facilitate the conditions (e.g., by contributing to higher levels of education and civic engagement) that can spur democratization. President Obama's speech on remittances in Cuba reflects this perspective.

---

[26] Some examples of these spending decisions include increasing government outlays in the lead up to an election (in democracies), co-opting political rivals with financial inducements (in non-democracies), and expanding a state's repressive capacity to fend off revolutionary threats.
[27] Tilly (1992). [28] Bueno de Mesquita and Smith (2010a); Ross (2013). [29] Bermeo (2016).
[30] Finkel et al. (2007); Bermeo (2011).
[31] Some examples include Finkel et al. (2007) and Bermeo (2011, 2016).

In democracies, findings about the link between remittances and political participation are mixed. Survey data from Mexico, for instance, suggest that remittance recipients are less likely to vote and seek out political information but are more likely to engage in local nonelectoral organizations.[32] Some organizations help migrant households pool their resources to finance various local public goods.[33] This self-provision of public goods can further erode a remittance recipient's desire to participate in politics.[34]

While survey evidence about the effects of remittances on political participation in autocracies remains scant, some scholars argue that remittances may influence political outcomes in electoral autocracies. Escriba-Folch et al. (2015), for instance, posit that remittance recipients are frequently less financially dependent on patronage networks that underlie the durability of many party-based dictatorships. As a consequence, remittance recipients are less likely to vote for the incumbent, increasing the prospect of democratization. Escriba-Folch et al. corroborate this conjecture cross-nationally. Their findings suggest that remittances can potentially undermine authoritarian stability, at least in party-based dictatorships. Of course, not all autocracies are party based or permit elections that can remove an incumbent.[35] Thus, these authors' findings may not be generalizable to autocracies with different institutional features.

Moreover, the study of Escriba-Folch et al. – like those of others in this nascent literature – faces additional challenges. First, they tend to study the impact of remittances on politics in isolation from other capital flows. Doing so can be problematic. For example, countries that receive substantial amounts of remittance income are also some of the top foreign aid recipients. This omission can undermine both the empirical *and* theoretical relationships between remittances and "politics."[36] Second, many studies tend to discount the potential for governments to respond strategically to remittance income. One such response could be through the composition and/or level of government spending. For instance, if migrant households spend some of their remittance income on health care (which many do), a government could respond by decreasing its own expenditure on health care and reallocating its spending

---

[32] For example, Goodman and Hiskey (2008); Perez-Armendariz and Crow (2010).
[33] Adida and Girod (2011).
[34] Other studies posit that for those who do participate in the political process, remittance income can influence an individual's voting calculus, especially in clientelistic electoral environments that characterize politics in Mexico and other Latin American countries (Pfutze 2012; Ahmed 2017a).
[35] Electoral manipulation features prominently in many "hybrid autocracies" as a means to safeguard the incumbent autocrat from losing power.
[36] An exception is Ahmed (2012), which advances a theory that foreign aid *and* remittance constitute "unearned foreign income" that an autocrat can leverage to fund strategies of political survival. This book builds on that framework by accounting for and modeling inflows of FDI (to firms).

elsewhere.[37] In addition to shifting the composition of government spending, higher levels of aggregate remittances could expand *total* government spending, for example, by reducing the cost of sovereign borrowing.[38] Specifying when and why a government might potentially act in these ways would provide a richer analytical lens to better understand how remittances can affect politics.

Foreign Direct Investment

Like remittance income, FDI also bypasses the government. That said, FDI can affect a government's "engagement" with both multinational corporations (MNCs) and domestic firms. Existing scholarship on this engagement can be categorized into three broad categories. One strand examines why countries have become more open to FDI, for instance, by analyzing public attitudes toward FDI and the sources of policies that attract FDI (e.g., domestic regulations, international treaties). These studies frequently build on theories of international trade (e.g., Heckscher–Ohlin and Stopler–Samuelson theorems) to predict who will favor or oppose FDI.[39] A second strand considers how potential host country politics and policies influence the choice of countries in which firms invest. A third strand evaluates the effect of FDI on governance, human rights, and inequality in host countries.[40] The second and third strands are particularly important for understanding the relationship between FDI and domestic politics in receiving countries, and especially the differences across autocracies and democracies.

In this context, "the single most-studied research question about the politics of FDI in the last two decades" is about the "types of political regimes MNCs prefer for the countries in which they invest."[41] While most scholars hold that democracies tend to receive more FDI, they disagree on why.[42] One argument posits that the electoral incentives of policymakers in democracies to not breach

---

[37] Alternatively, while remittance income can be difficult for governments to track and tax directly at the source, a government could tax goods that migrant households may be more predisposed to consuming, such as investments in "better quality" education (e.g., private schools) and health care services (e.g., more expensive treatments).

[38] Singer (2012) finds that remittances both lower a country's default risk (by lowering its sovereign debt spread) and increase its government consumption. This book evaluates the veracity of these potential public finance channels, especially for governments in nondemocracies.

[39] For example, Pandya (2010) and Pinto (2013).

[40] For an overview on the political economy of FDI, especially for the first and second strands, see Pandya (2016).

[41] Pandya (2016), 462.

[42] For example, across a series of cross-sectional and panel regressions for a sample of more than 100 countries from 1970 to 1997, Jensen (2003, 605) concludes: "The most conservative long-run estimate of the effect of democracy on FDI inflows ... predicts a democratic country will receive an added 0.61% as a percentage of GDP which amounts to an increase of over 45% of FDI inflows." For a more skeptical view on whether democracies attract more FDI, see Li et al. (2018).

contracts better aligns with those of MNCs (e.g., these firms do not want their assets expropriated), while another hones in on the importance of property rights in lowering political risk. On the former, Jensen (2003), for example, advances an accountability mechanism in which voters in democracies punish elected leaders who breach foreign contracts – a commitment device that lowers the political risk faced by MNCs investing abroad. On the latter, Li and Resnick (2003) posit that democracy is a double-edged sword for MNCs evaluating political risk. Democracies tend to better safeguard property rights, but policymakers also face greater electoral pressures to expropriate from MNCs to support greater redistribution to voters. Autocracies, by contrast, can exercise greater autonomy in trading off weaker property rights protection with the state's ability to cater to MNCs, including the provision of monopoly access to the local market and investment incentives. Relatedly, scholars have probed additional possible explanations for the seeming "democratic advantage" in FDI inflows, including constraints faced by the country's chief executive, the strength of the judiciary in enforcing contracts, policy fragmentation, and the partisanship of governments.[43]

Surprisingly, studies examining the *reverse* question of the impact of FDI on regime type have received relatively less scholarly attention. This is curious because the pernicious effects of multinational firms on governance features prominently in accounts of globalization during colonialism.[44] The potential for leveraging FDI to buttress nondemocratic rule has not escaped the governing strategies of some dictators in the postcolonial era. For instance, on coming to power in the early 1960s, military governments in Indonesia and Brazil assigned a large role to FDI in their five-year development plans. And in Egypt, joint ventures with foreign investors continue to underlie the strong influence of the military in Egyptian politics.[45] For these dictatorships, foreign investors create opportunities for rent extraction. They can also generate downstream economic growth, thereby reinforcing the regime's legitimacy among the general public.

---

[43] Li (2009); Staats and Biglaiser (2012); Henisz (2000); Pinto (2013).

[44] For example, joint-stock companies played an important role in expanding and consolidating British colonial rule. In exchange for monopoly rights, commercial enterprises were granted political authority in governing England's overseas colonies in Canada, India, and Africa (Rodrik 2011). In many instances, these companies governed oppressively by denying political rights and pitting ethnic groups against each other. Other colonial powers also "invested" abroad with devastating effects. King Leopold of Belgium, for instance, decimated the native population in the Congo in order to export rubber and reap all the proceeds for his own use (Hochschild 1998).

[45] On Indonesia and Brazil, see Robison (1986) and Evans (1979), respectively. Marshall and Stacher (2012) describe how FDI has been one of the main channels through which the Egyptian military has expanded its influence on both national and local economies. In particular, the military has forged joint ventures with MNCs in the transport and energy sectors under the support of the Egyptian environment. In the maritime industry, for instance, the Egyptian military partnered with the world's largest MNCs. In these projects, the military maintained considerable shares and executive positions through a state-owned holding company.

1.2 Existing Research

While the insights from these qualitative studies have been examined more recently in other autocracies (e.g., China, Vietnam), often with sophisticated statistical approaches, there are surprisingly few cross-national studies investigating whether FDI affects the stability of autocracies.[46] Rather, such research examines the relationships between FDI and correlates of regime type, such as the quality of human rights in host countries.[47] To the extent this scholarship truly hones in on the impact of FDI on authoritarian politics, it tends to do so in isolation of other capital flows, such as foreign aid and remittances.[48]

This omission frequently occurs in both the underlying theory and empirics linking FDI to politics in host countries. On the former, existing research tends to downplay the agency of governments in responding strategically to inward FDI. For example, what policy levers might a government employ to exploit FDI to its own ends? And how do these levers differ by host country regime characteristics and/or by the sectors foreigners opt to invest in? Empirically, concerns with reverse causality and other forms of endogeneity bias loom large in evaluating the *causal* impact of FDI on politics.

Sovereign Borrowing

While foreign aid, remittances, and FDI are important sources of capital in developing countries, another relatively "sticky" type of foreign capital is sovereign debt.[49] As a source of funds that accrue directly to governments, sovereign borrowing can generate political ramifications similar to those of foreign aid.[50] Sovereign borrowing can finance repression and patronage. Yet, sovereign debt differs from aid, remittances, and FDI in at least two important ways.

---

[46] For example, observing that corruption is often a means of "rewarding" regime supporters, Zhu (2016) shows that Chinese provinces with greater MNC activity foster greater corruption among provincial-level government officials. In Vietnam, Malesky (2008) documents that FDI can empower subnational leaders with greater autonomy over economic policy, thereby improving their effectiveness and, in turn, strengthening the dictatorship's legitimacy.

[47] Cingranelli and Richards (1999); Apodaca (2002); Blanton and Blanton (2007).

[48] A few scholars have documented a correlation between these capital flows. For example, Leblang (2010) finds that countries linked by larger migrant networks (and remittance inflows) generate more investment into the migrants' country of origin. Foreign aid may also spur inward FDI. Aid can "reveal" private information from donors or lenders about the relevant host country features, or it can create external constraints on policymakers that can lower the political risk for foreign investors (Jensen 2004; Biglaiser and DeRouen 2010).

[49] On the political economy of sovereign debt, see Lipson (1985), Frieden (1991), and Tomz (2007).

[50] Chapter 7 discusses how sovereign debt can be integrated into the book's theory. Moreover, scholars have examined how sovereign borrowing can empower dictators (e.g., DiGiuseppe and Shea 2016).

First, while governments in developing countries do borrow from abroad, the bulk of sovereign debt is still issued by governments in developed countries.[51] In contrast, inflows of foreign aid and remittances are overwhelmingly concentrated in developing countries, and about half of all inward FDI now resides outside advanced economies.[52] Second, the ability of governments in developing countries to borrow is *less* affected by "exogenous" factors compared to their capacity to attract foreign aid, remittances, and FDI. A government's monetary and fiscal policy decisions (e.g., interest rate, composition of public spending) greatly affect sovereign borrowing in developing countries.[53] In contrast, the decisions of donor governments affect their aid disbursements, while economic conditions in "sending" countries influence flows of remittances and FDI into "receiving" countries. The greater exogeneity of these capital flows means recipient governments may need to employ different strategies to benefit from them. As this book argues, governments can treat foreign capital as nontax income to finance politically beneficial policies.

1.3 CONTRIBUTIONS

This book offers a unified theory of the impact of these three prominent types of international capital on the survival of authoritarian regimes. Existing literature that examines different types of international capital flow in isolation misestimates their effects. The theoretical approach in this book clarifies the channels through which a strategically oriented government can leverage each type of capital flow to finance two important instruments of authoritarian survival: repression and patronage. The methodological approach of this project takes seriously the questions of causal identification, exploiting exogenous variation in foreign capital to more precisely estimate the magnitude of the effects of capital flows. Ultimately, the book concludes that international capital fosters authoritarian politics. The results of this study challenge many existing studies and contribute to important literatures in economics and political science.

Globalization and Democracy

Since at least the time of Immanuel Kant, many political theorists – and a large swath of political economists – have viewed greater international economic exchange as a means of fostering democracy and promoting "perpetual peace" among sovereign states.[54] In regard to perpetual peace, a voluminous

---

[51] Guzman et al. (2016).
[52] About 49.3 percent of FDI has been invested in countries outside of the Organisation for Economic Development and Co-operation (OECD) since 2010 (OECD 2018).
[53] Mosley (2003).   [54] Kant (1795).

1.3 Contributions

literature exists on economic interdependence and war, most recently manifesting in lively academic debates pitting the veracity of democratic peace theory against that of the capitalist peace.[55] Concerning connections to democracy, contemporary scholars have probed the theoretic and empirical relationships between globalization and democracy.[56] One strand of this scholarship investigates whether democracy propels trade liberalization, while another evaluates the conditions under which globalization strengthens (or undermines) democratic governance.[57]

While much of this literature focuses on international trade, studies also examine – albeit to a lesser degree – other aspects of globalization, such as the cross-border movement of capital and labor, and the resultant impacts on democratic *and* nondemocratic politics. This book deals most directly with this second line of inquiry, scholarship, as noted already, that tends to evaluate each type of cross-border capital flow in isolation and treats governments as relatively passive agents. This book addresses these issues by presenting a theory and testing whether governments can exploit inflows of foreign aid, remittances, and FDI to their political advantage. In doing so, it raises skepticism over Kant's assertion that globalization enhances democracy.

Authoritarian Survival

The evidence I present contributes to a rich literature in comparative politics about the determinants of democratic and authoritarian politics.[58] Recent scholarship provides stronger microfoundations for the strategies dictators use to ensure their survival, including reducing the likelihood of military coups, identifying and co-opting political rivals through "hybrid institutions" (e.g., elections in autocracies), and gaining support from powerful foreign patrons.[59] As this scholarship recognizes, a dictator chooses the strategy that will minimize his institutional constraints and maximize his sources of revenue.

---

[55] For instance, on interdependence and war, see Hirschman (1945) and Copeland (2014). See Gartzke (2007) for a solid treatment of the democratic versus capital peace debate.

[56] For an overview of this scholarship, see Eichengreen and Leblang (2008) and Milner and Mukherjee (2009). Rudra (2005) argues that trade and investment enhance democracy in countries if "safety nets" are used simultaneously as a strategy for providing stability and building political support.

[57] For example, Milner and Kubota (2005) argue that democratization has propelled trade liberalization in developing countries since the 1970s. In contrast, Rodrik (2011) emphasizes how globalization can challenge policy autonomy and principles of democratic governance in both developed and developing countries.

[58] Some important works include Moore (1966), Huntington (1968), O'Donnell (1973), Wintrobe (1998), Przeworski et al. (2000), Bueno de Mesquita et al. (2003), and Acemoglu and Robinson (2006). This list is *far* from exhaustive.

[59] On microfoundations, see Wintrobe (1998) and Svolik (2012). On hybrid political institutions in autocracies, see Ghandi (2008). Jamal (2012) provides an excellent account of how foreign governments ("patrons") buttress autocracies in the Middle East.

This book contributes a new international financial dimension to the abundant existing literature. Comparative politics emphasizes the domestic sources of political survival and regards the international economy as largely exogenous to the decision-making of government leaders. This book shows that governments use foreign capital to secure their incumbency, such as channeling rents from inward FDI to the military (thus, lowering the odds of a military coup), and allocating revenues from foreign aid and remittances to buy off political rivals. Given the tremendous growth of foreign aid, remittances, and FDI across the globe, this book's insights are therefore applicable across geographic regions and demonstrate the benefits of integrating research in international political economy (IPE) and comparative politics (CP).

Resource Curse

The theoretical approach of this integration of IPE and CP regards international capital as a type of nontax income in much the same way a government can exploit natural resources (e.g., oil, gas, diamonds, rubber, copper). In many instances, nontax income can be a source of political stability in both democracies and autocracies, at the cost of lowering economic growth and financing repression and patronage.[60] This "political resource curse" is often magnified in autocracies, a conclusion that is also applicable to the political consequences of capital inflows.

In Chapter 3, I develop a theory arguing how a country's political institutions shape its government's response to inflows of foreign aid, remittances, and FDI. I argue that a greater fraction of foreign aid is siphoned off to finance patronage in autocracies than in democracies. And in autocracies, a government responds to remittance income – which accrues to households – by reducing a greater share of its expenditures on welfare goods and redirecting those funds to patronage. The income and substitution effects are magnified in autocracies. In short, the willingness and ability of autocrats to treat international capital like nontax income generates antidemocratic effects analogous to those of the resource curse.

International Development and Contemporary Politics

The evidence I use to test whether international capital enhances authoritarian politics spans the entire globe. Chapter 4 shows that aid disbursed by the

---

[60] On the politically stabilizing effects engendered from nontax income, see Morrison (2015). On the nefarious political consequences (e.g., conflict, repression) associated with natural resources, see Ross (2012, 2013). On the pernicious effects associated with foreign aid, see Bueno de Mesquita and Smith (2010b). Dunning (2008), however, challenges the view that natural resources always foster autocracy. Menaldo (2016) advances a conditional theory based on an "institutions curse" linking natural resources to economic development.

1.3 Contributions

world's largest bilateral donor – namely the United States – allows recipient governments to repress their populations. Because political rights spur economic development, the evidence I present is salient for practitioners in international development.[61] Moreover, as geostrategic interests greatly influence US aid disbursements, this may further explain why many US allies remain nondemocratic despite pronouncements from the US government that its economic assistance strives to promote freedom and democracy abroad.

The other types of capital flows can also engender antidemocratic effects. For instance, FDI in countries with high fixed cost industries (e.g., oil) can create opportunities for patronage that rewards regime supporters, an assertion confirmed by the empirical analysis in Chapter 6. This finding explains the resilience of several military dictatorships during the 1960s and 1970s that actively courted FDI (e.g., Brazil, Indonesia), as well some contemporary autocracies that reward regime members by extracting rents from FDI projects (e.g., China, Egypt).

With respect to remittances, Chapter 5 provides compelling evidence that remittances have helped consolidate dictatorial rule in many Muslim majority countries since the 1970s, a finding that offers a new interpretation for the "democratic deficit" in many Muslim societies. The comparative vignette in Chapter 5 shows that sufficiently buoyant remittance income during the Arab Spring allowed Jordan's King Hussein to retain the loyalty of regime supporters (through the distribution of patronage) and thus mollify any attempt at his removal from power. This was not the case in Egypt, where a low amount of remittance income accelerated the demise of the Mubarak dictatorship in 2011.

Causal Inference

Evaluating the causal effect of international capital on politics with observational, real-world data can be problematic. One of the principal challenges is endogeneity: capital inflows are correlated with domestic politics in recipient countries and vice versa. For instance, donors may disburse foreign aid to accelerate political reform in democracies, in which case countries with "good institutions" (less autocratic) may be "rewarded" with higher amounts of aid. If this is the case, estimating the effect of aid on measures of autocracy in a regression framework will be "biased." Overcoming this challenge requires identifying an exogenous source of variation for the endogenous variable (e.g., aid in the foregoing example) that is uncorrelated with the outcome variables of interest (e.g., political institutions, political tenure, government expenditures, corruption). This exogenous variable (Z) can then be used to "explain" some of the variation in the endogenous variable (X), which in turn then can be used to explain some of the variation in the key outcome variable (Y). This two-stage process can be modeled in a regression framework that can

---

[61] According to Sen (1999), political freedoms constitute and propel economic development.

generate confidence intervals for the coefficient estimates and control for potential confounders.[62]

In practice, the quest to "identify" plausibly exogenous sources of variation that meet the assumptions for valid two-stage estimation is challenging.[63] However, rather than abandon observational data, this book introduces creative ways to turn the observable world into a quasi-experimental laboratory. To do so, I identify quasi-random variables that correlate with the capital flow of interest. For example, I show how the composition of the US Congress – and its interaction with the probability that a country receives US aid – correlates with US bilateral foreign aid but is unrelated to domestic economic and political conditions in recipient countries. In a separate analysis, I show how world oil prices – interacted with a Muslim country's distance to Mecca – correlates with remittances, but again is unrelated to domestic economic and political conditions in recipient countries. In a different setting, I use the inherently "hit or miss" nature of oil discoveries to generate a stochastic pattern of FDI increases that is uncorrelated with domestic politics in recipients. By relying on these quasi-random determinants of each capital flow, I am able to identify and estimate the causal effect of each type of capital on many dependent variables in recipient countries. Throughout the analysis, I am careful to articulate the scope, conditions, and generalizability of these causal inferences.

1.4 THE PATH FORWARD

The book proceeds as follows. Chapter 2 refines the book's scope of inquiry and demonstrates the main claim in the *raw data*: that international capital empowers dictators, but not democrats. The chapter has two main objectives. First, it defines foreign aid, remittances, and FDI and describes some of the spatial and temporal variation in international capital and authoritarianism. Second, the chapter provides preliminary evidence that international capital is positively associated with longer leader tenure in autocracies, but not in democracies.[64] This association raises two important questions. The first is theoretical: why should international capital help a dictator's political fortunes but not that of democrats? The second is empirical: given that capital flows and "politics" are potentially endogenous, how can one gauge the causal impact of foreign aid, remittances, and FDI on authoritarian politics?

---

[62] In applied social science research, this two-stage process is often called two-stage least squares (2SLS) or instrumental variable (IV) regression. For more on its origin, theory, and applications, see Stock and Trebbi (2003) and Angrist and Pischke (2009).
[63] Sovey and Green (2011) provide a useful "reader's guide" to properly executing two-stage models.
[64] The chapter shows that international capital is positively associated with leader tenure in *increasingly* authoritarian polities.

1.4 The Path Forward

Chapter 3 addresses the first question. It develops a revenue-based theory that explains how governments can harness international capital to finance strategies of political survival. These strategies can affect both the quality of *governance* (e.g., corruption, composition of public spending) and *institutions* (e.g., constraints on the executive). In particular, international capital exhibits nontax properties that can fund two important strategies of political survival in autocracies: repression and patronage. In nondemocracies, this patronage can be in the form of higher military spending, corruption, and compensation for public sector employees. Each capital inflow can finance repression and patronage through a distinct channel: an income effect associated with foreign aid, a substitution effect associated with remittances, and a rentier effect associated with FDI.

Determining whether capital inflows *cause* authoritarianism is challenging. The principal worry is endogeneity. One source of this bias is reverse causality: political conditions in recipient countries (e.g., quality of political institutions, property rights, fate of leaders) affect capital inflows.[65] Solving this problem requires "identifying" a *plausibly exogenous* source of variation for the endogenous variable (i.e., the relevant type of international capital) that is also uncorrelated with the outcome variables of interest (i.e., political institutions, political tenure, government expenditures, corruption, etc.). As the theory identifies a *distinct* channel for each capital flow, I devote a separate chapter for *each* type of capital to tackle this endogeneity problem. In the empirical chapters to follow, I focus my attention on developing countries that are the main recipients of foreign aid, remittances, and FDI.[66] Table A1 lists this sample of developing countries.

Chapter 4 evaluates the impact of foreign aid from the world's largest bilateral donor – the United States – on the quality of democracy in its aid recipients. A case vignette first describes how US foreign aid financed the repressive clan-based rule of General Siad Barre in Somalia over two decades. The core of the chapter is the cross-national statistical analysis of US economic aid disbursements. To mitigate concerns associated with endogeneity, I develop an instrumental variable (IV) for US bilateral economic aid disbursements. As described in greater detail in Chapter 4, the research design builds on the institutional foundations of US aid decisions, in

---

[65] For instance, donors may disburse foreign aid to accelerate political reform in democracies (e.g., Finkel et al. 2007), in which case countries with "good institutions" (less autocratic) may receive higher amounts of aid. Some donors, however, may also use aid to "buy" policy concessions, especially from autocracies where the "cost" of doing so is lower because of a smaller winning coalition to buy off (Bueno de Mesquita and Smith 2009). Thus, donors may disburse higher amounts to less democratic countries. In addition to reverse causality, omitted variable bias and measurement error can be sources of endogeneity bias.

[66] Countries that are not recipients of *all* three types of capital flows are excluded from the empirical analysis. For example, the United States and United Kingdom are excluded because they are not recipients of foreign aid. Rather, each country is an important foreign aid donor.

which the funding and allocation of bilateral economic aid involves both the executive branch and Congress. Congress – and its composition – determines the aid budget.

In years of greater legislative fragmentation, Congress tends to spend more on the total aid budget. Because the composition changes every two years – primarily because of district-level characteristics – legislative fragmentation is plausibly exogenous to prevailing economic and political conditions in US aid recipients. Building on these insights, I exploit plausibly exogenous variation in the legislative fragmentation of the US House of Representatives, and interact this with the probability that a country receives US aid to construct a powerful instrumental variable for US bilateral aid.[67]

Armed with this instrumental variable, the results demonstrate that US economic aid harms political rights, expands the powers of leaders, and makes authoritarian institutions more durable. To unpack the underlying channel, I provide stronger evidence that US aid harms political rights by reducing a recipient government's "tax effort" rather than dramatically increasing its repressive capacity. The former is consistent with existing research that identifies a trade-off between government accountability and taxation: a government that exerts less tax effort (i.e., collects a lower share of its revenues from taxation) tends to be less politically accountable to its population (i.e., citizens enjoy fewer political freedoms).[68]

In a similar fashion, Chapter 5 presents cross-national evidence that remittance income can also embody nontax properties such as foreign aid. The chapter demonstrates that remittances can finance authoritarian politics via a substitution effect: remittances can allow leaders to reduce their provision of welfare goods (to the masses, including intended remittance recipients) and divert those unused resources to fund patronage. To overcome challenges posed by the endogeneity of remittances with authoritarian politics – stemming primarily from measurement error and reverse causality – I leverage a novel quasi-natural experiment of oil price–driven remittance flows emanating from the Persian Gulf to non–oil-producing Muslim majority countries in North Africa, the Middle East, and South Asia.[69]

During periods of high oil prices, Gulf oil producers tend to import labor from non–oil producing Muslim countries (e.g., Bangladesh, Pakistan), particularly those geographically closer to these oil producers (e.g., Jordan, Lebanon). These migrants in turn remit large amounts of their earnings to their home countries

[67] This "probability" captures how temporal changes in legislative fragmentation are propagated to different aid recipients.
[68] See Tilly (1992) on this trade-off between taxation and political accountability.
[69] Concerning the former, poorer and autocratic countries, for instance, may systematically misrecord remittance inflows. Regarding the latter, political conditions in a "recipient country" can affect emigration and subsequent remittance inflows. The "endogeneity problem" associated with remittances is described in greater detail in Chapter 5.

1.4 The Path Forward

(many of whom tend to be nondemocratic).[70] For these labor-exporting countries, their aggregate remittances are thus correlated with the world price of oil and their geographic distance to the Persian Gulf. In particular, the exogeneity of world oil prices forms the basis of a time-varying instrumental variable for remittance income. The instrument interacts the world price of oil with a recipient country's distance to Mecca.[71] Using this instrument, I show that remittances cause autocratic leaders to face lower constraints on their power, extend their duration in office, and make their country's political institutions less democratic. I also present evidence consistent with the underlying channel: that remittances reduce a government's expenditures on welfare goods in favor of increased spending on public sector compensation and corruption (both proxies for patronage).[72]

This empirical finding that remittances can be a source of authoritarian politics offers a new perspective in understanding the divergent political dynamics of Egypt and Jordan during the 2011 Arab Spring. In a "comparative vignette," I argue that buoyant amounts of remittances provided sufficient rents to sustain patronage in Jordan and thus hindered the ousting of the ruling monarch during the Arab Spring. This was not the case in Egypt, where declining remittance income may have weakened the Mubarak regime's ability to finance its various patronage networks (e.g., the military), thus making the regime more vulnerable to removal from office.

Of course, attracting foreign aid and remittances may not be the most attractive form of international capital for all dictators. For autocratic governments determined to industrialize their countries (such as those in Latin America, East/Southeast Asia), attracting sufficient FDI – especially in industries with high fixed costs and capital requirements – has been instrumental in economic development. This economic development, in turn, has served to legitimize the state's authoritarian rule and has generated rents for distribution to regime supporters. This notion underlies the analysis in Chapter 6.

Chapter 6 presents cross-national evidence that FDI, particularly in high fixed cost industries (e.g., oil exploration, petrochemicals), can create rents that an authoritarian government can use to fund the military. In doing so, these governments are also able to retain the loyalty of the military, a key domestic ally. To gauge the causal effect of FDI on authoritarian politics, I examine how FDI inflows following the random discovery of oil affect a government's military expenditures. The empirical strategy is a two-step process. I first leverage plausibly exogenous variation in oil discoveries to

---

[70] Chapter 5 carefully argues and shows that the average treatment effect (ATE) of remittances on politics pertains to a sample of predominantly nondemocratic Muslim-majority non–oil-producing countries.
[71] A country's distance to Mecca proxies for the fixed cost of migration to the Persian Gulf.
[72] See Keefer (2007) on using public sector employment as a measure of patronage in autocracies.

evaluate how *new* opportunities for rents lead to higher inflows of FDI. I show that new oil discoveries lead to inflows of FDI in subsequent years. I then show how this oil-discovery–induced FDI inflow causes higher levels of military spending in autocracies, but not in democracies.

Moreover, in addition to generating rents for the military, attracting FDI is part of a broader strategy for many dictatorships in strengthening their national security and fostering broad-based economic growth. To further trace the causal process, I describe how attracting foreign capital played an integral part in General Suharto's dictatorial rule in Indonesia for more than three decades. Attracting FDI proved successful in generating rents for Suharto's inner circle of supporters (i.e., the military, his family) and spurred broader economic growth. However, the regime's response to the 1997 Asian Financial Crisis contributed to a mass exodus of FDI. This capital outflow, in turn, accelerated the demise of Suharto's reign and facilitated Indonesia's subsequent transition to democracy. This last point highlights the dangers of capital outflows: while foreign capital inflows can empower autocrats through the generation of rents, a reversal can potentially cripple those dictatorships.

Chapter 7 concludes. It summarizes the book's argument and evidence. The chapter also discusses a number of potential extensions to both the underlying theory and empirical testing, as well as some economic and political welfare implications of foreign capital.

Appendix

TABLE A1 *Sample of developing countries*

Afghanistan	Czech Rep.	Lebanon	Saudi Arabia
Albania	Djibouti	Lesotho	Senegal
Algeria	Egypt	Liberia	Serbia
Angola	El Salvador	Libya	Sierra Leone
Argentina	Equatorial Guinea	Lithuania	Slovak Rep.
Armenia	Eritrea	Madagascar	Slovenia
Azerbaijan	Estonia	Malawi	Solomon Islands
Bangladesh	Ethiopia	Malaysia	South Africa
Belarus	Fiji	Mali	South Sudan
Benin	Gabon	Mauritania	Sudan
Bhutan	Gambia	Mauritius	Suriname
Bolivia	Ghana	Mexico	Swaziland
Bosnia & Herzegovina	Guatemala	Moldova	Tajikistan
Botswana	Guinea	Mongolia	Tanzania
Brazil	Guinea-Bissau	Montenegro	Thailand
Bulgaria	Guyana	Morocco	Timor-Leste
Burkina Faso	Haiti	Mozambique	Togo
Burundi	Honduras	Myanmar	Trinidad & Tobago
Cabo Verde	Hungary	Namibia	Tunisia
Cambodia	India	Nepal	Turkey
Cameroon	Indonesia	Nicaragua	Turkmenistan
Central African Rep.	Iran	Niger	Uganda
Chad	Iraq	Nigeria	Ukraine
Chile	Israel	Oman	Uruguay
China	Jamaica	Pakistan	Uzbekistan
Colombia	Jordan	Panama	Venezuela
Comoros	Kazakhstan	Papua New Guinea	Vietnam
Congo, Dem. Rep.	Kenya	Paraguay	Yemen, Rep.
Congo, Rep.	Korea, Rep.	Peru	Zambia
Costa Rica	Kosovo	Poland	Zimbabwe
Cote d'Ivoire	Kyrgyz Rep.	Romania	
Croatia	Lao PDR	Russia	
Cyprus	Latvia	Rwanda	

Note: The table shows a sample of developing countries that receive nonnegative amounts of foreign aid, remittances, *and* foreign direct investment.

2

International Capital and Authoritarian Survival
A Descriptive Overview

This chapter has two objectives. First, I define and describe the spatial and temporal variation in international capital and authoritarian politics.[1] The former comprises the book's key independent variable, while the latter is the book's main dependent variable. Second, I illustrate the existence of an *association* between international capital and political survival in autocracies in the *raw data*. Such an examination is useful as it illustrates the book's central empirical prediction without having to "finesse" the data to establish a statistical relationship between international capital and authoritarian politics. These associations provide motivation for further theoretical and empirical inquiry. To be clear, I do *not* claim a causal relationship between international capital and authoritarian politics in this chapter. Causal evaluations are the focus of Chapters 4 to 6.

Section 2.1 addresses the first objective. It narrows the scope of theoretical and empirical inquiry that the rest of the book tackles in greater detail: the channels through which international capital can foster authoritarian politics. In a nutshell, international capital is composed of cross-border financial transfers between national governments (foreign aid), individuals (remittances), and firms (FDI).[2] Authoritarian politics is conceptualized on two important dimensions: its institutional features and the political survival of leaders.

Section 2.2 addresses the second objective. Starting with a series of bivariate plots of the raw data, I first illustrate a positive association between international capital flows and leader tenure in countries with more entrenched authoritarian institutions. I corroborate this finding with more stringent analyses that scholars employ to evaluate the determinants of political survival. Controlling for potential

---

[1] "Politics" is multifaceted. Here it refers to a country's system of *institutions* and the *political survival* of its leaders. This chapter examines the association between international capital and political survival across countries with different institutional settings.
[2] As discussed in Chapter 1, I do not focus on sovereign debt.

confounders, such as measures of economic development and unobservable country and temporal effects (with country and year fixed effects), I show that greater inflows of international capital lower the likelihood of leader termination in autocracies but *not* in democracies. Moreover, this association is statistically *more robust* in countries with *more* authoritarian institutions.

Building on these preliminary relationships, Section 2.3 concludes the chapter by discussing the next steps to establish a *causal* relationship – both theoretical and empirical – between international capital and authoritarian politics.

2.1 MEASUREMENT

Defining International Capital

In the broadest sense, international capital includes any transfer of financial assets across national borders – i.e., the cross-border transfer of funds between national governments, individuals, and firms, and their combination.[3] This book focuses on three important types of international capital flows involving three distinct actors: foreign aid between *governments*, remittances between *households* (individuals), and foreign direct investment between private *firms*. Narrowing the theoretical scope of international capital to these three types is appropriate, as these cross-border transfers are salient features of the global economy. Combined, these transfers have grown worldwide from around $120 billion in 1970 to about $2.4 trillion by 2015, and for many countries, capital inflows are quite large, comprising more than 25 percent of their national economies (e.g., Armenia, Honduras, Liberia, Jordan, Moldova, Mozambique).[4] Individually, each capital flow is unique, as it involves distinct actors and is often "received" in different economic and political contexts.

Foreign aid is the international transfer of capital, goods, or service from a donor country or international organization to a recipient country.[5] The transfer is typically from a wealthy donor to a poor recipient.[6] Foreign aid

---

[3] This also includes the transfer of funds involving multilateral organizations comprising national governments, such as the United Nations, World Bank, International Monetary, European Union, etc.

[4] Country averages (in constant 2010 US$) are calculated from 2000 to 2015. Data are from the World Bank and the author's calculations. In 2008, international capital peaked globally to $3.85 trillion. International capital flows declined significantly in the aftermath of the 2007–2008 global financial crisis. International capital flows have grown steadily since 2010.

[5] Aid can also "benefit" the donor country. As an instrument of a donor's foreign economic policy (Alesina and Dollar 2000), foreign aid can advance geopolitical goals (e.g., gain allies during the Cold War) and disbursed to "buy" policy concessions from recipient countries (Bueno de Mesquita and Smith 2009).

[6] The top bilateral aid donors (in absolute dollar amounts) are the United States, Japan, Germany, and the United Kingdom. Regional development banks, the United Nations, and the World Bank are some of the leading multilateral organizations that disburse foreign aid.

can involve the transfer of financial resources, goods (e.g., food) or technical advice and training. In many instances, the resources can take the form of grants or concessional credits, such as export credits. In practice, the most common type of foreign aid is official development assistance (ODA), which aims to promote development and combat poverty.[7] The primary source of ODA is bilateral grants from one government to another government.[8]

While foreign aid transfers resources between governments, migrant remittances do not involve governments directly.[9] Rather, a remittance is a transfer of money by a foreign worker to an individual (frequently a family member) in his or her home country. Remittance income therefore requires emigration to another country – either temporarily (e.g., guest workers) or permanently (e.g., gaining citizenship in the "host" country).

Measuring remittance flows is often difficult because they may flow through unofficial channels (e.g., Hawala in many Islamic countries), and governments in many recipient countries lack the capacity to accurately track and record these capital inflows.[10] As the cross-border transfer of money involves different currencies (and hence a currency conversion), a country's central bank will record official remittance inflows and outflows. These figures are then reported to the World Bank and the IMF, which underlies cross-national analysis of remittance data.[11]

---

[7] Official development aid is defined by the Organisation for Economic Co-operation and Development.

[8] In practice, some aid may be in the form of loans, and sometimes the aid is channeled through international organizations (e.g., the International Monetary Fund [IMF], World Bank) and nongovernment organizations (CARE, Oxfam).

[9] At the aggregate level, foreign aid and remittance income enter a country's balance of payments. As a capital inflow, both foreign aid and remittances are a net current transfer and are recorded in a country's capital account. For recipient ("home") countries a positive inflow is a credit, while for sending ("host") countries, the outflow is a debit.

[10] Hawala is a method of transferring money without any actual movement. It is an alternative remittance channel that exists outside traditional banking systems. Transactions between Hawala brokers are done without promissory notes because the system is heavily based on trust.

[11] Most applied cross-national research employs data from the World Bank (which is based on information from the IMF). The World Bank measures "personal remittances," which is the sum of personal transfers and compensation of employees. Personal transfers are a broader definition of worker remittances. Personal transfers include all current transfers in cash or in kind between resident and nonresident individuals, independent of the source of income of the sender (and regardless of whether the sender receives income from labor, entrepreneurial or property income, social benefits, or any other types of transfers; or disposes assets) and the relationship between the households (regardless of whether they are related or unrelated individuals). Compensation of employees refers to the income of border, seasonal, and other short-term workers who are employed in an economy where they are not resident and of residents employed by nonresident entities. Compensation of employees represents remuneration in return for the labor input to the production process contributed by an individual in an employer–employee relationship with the enterprise. Compensation of employees has three main components: wages and salaries in cash, wages and salaries in kind, and employers' social contributions. Compensation of employees is

2.1 Measurement

Like remittance income, foreign direct investment (FDI) does not accrue directly to governments.[12] Rather, FDI entails the transfer of financial assets – and potentially other firm-specific assets, such as managerial and technological expertise – across firms in different countries. Its principal aim is to either establish business operations or acquire business assets, such as ownership or a controlling interest in the host country firm.[13] Moreover, unlike foreign aid, which is typically a financial transfer from a wealthy donor to a poorer recipient, FDI can flow between firms in developed (rich) *and* developing countries. While the bulk of FDI involves firms in developed countries (i.e., "North–North"), it increasingly involves investments *originating* from firms in developing countries.[14]

Variation in International Capital

Figure 1.1 in Chapter 1 depicts the tremendous growth in foreign aid, remittances, and FDI since the 1970s, and Figure 1.2 describes some of the spatial variation in international capital across the world. Each figure captures some degree of the "aggregate" variation in international capital. In this subsection, I provide additional description of this variation across *time*, *space*, and *composition*. This description also quantifies the economic significance of these capital flows.

Temporal Variation. International capital flows have grown substantially since the late 1970s.[15] Figure 2.1 plots the annual global aggregate of each capital flow (in log units) since 1970. In the mid-1970s, governments disbursed about $73 billion in foreign aid, migrant workers remitted about $40 billion globally and firms invested about $90 billion in other countries.[16] These flows have steadily increased over time. By 2010, donors had nearly doubled their annual disbursements of aid worldwide to about $130 billion, while the worldwide

---

recorded gross and includes amounts paid by the employee as taxes or for other purposes in the economy where the work is performed.

[12] FDI can have an indirect effect on governments. For instance, it can affect government behavior, such as corruption (e.g., Malesky et al. 2015; Pinto and Zhu 2016). Moreover, because FDI can be a source of economic growth, many governments strive to attract foreign investment (e.g., Buthe and Milner 2008).

[13] Indeed, the intent of the investment to establish either control of, or at have at least substantial influence over, the decision-making of a foreign business distinguishes it from other types of international investments, such as portfolio investments in which an investor merely acquires equities of foreign-based companies.

[14] United Nations (2006).

[15] Reliable and comparable cross-national data on foreign aid, remittances, and FDI is available only from 1970 onwards.

[16] Calculations based on data from the World Development Indicators. All values in 2010 US dollars.

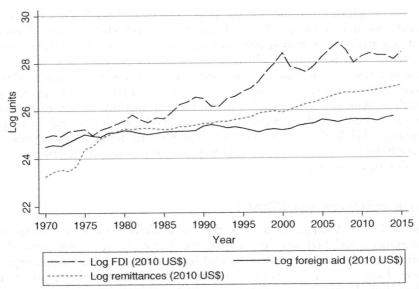

FIGURE 2.1 Total foreign aid, remittances, and FDI worldwide, 1970–2015.
Source: World Development Indicators and the author's calculations.

totals of remittances and FDI had surged more than tenfold from their respective levels in 1975.[17] Annual flows of FDI have consistently exceeded those of foreign aid and remittances (as evident in Figure 2.1), and the annual growth of FDI has been more volatile than that associated with foreign aid and remittances.[18]

Spatial Variation. The recipients of foreign aid, remittances, and FDI vary geographically (spatially). Using data from the World Bank, Table 2.1 lists the top ten recipients of each capital flow since 2000. In each row, the numeric value in each cell reports the corresponding country's average intensity of each capital flow between 2000 and 2015. This intensity is measured as the share of each capital flow relative to its country's national income. For example, foreign aid has, on average, comprised nearly half (approximately 48 percent) of Liberia's national income since 2000. In general, countries in Africa and the "greater" Middle East (e.g., West Bank and Gaza, Afghanistan) round out the top aid recipients. In many

---

[17] In 2010 US$, remittances and FDI worldwide totaled about $420 billion and $1.9 trillion, respectively.
[18] A simple measure of volatility is the standard deviation in the annual growth rate of each capital flow. Using this metric, the standard deviation in the annual growth rates of aggregate FDI (24.1 percent) exceeds those of aggregate aid and remittance inflows (8.3 and 13.4 percent, respectively).

2.1 Measurement

TABLE 2.1 *Top recipients (as a percentage of GDP) of foreign aid, remittances, and FDI: Country average over the 2000–2015 period*

Country	Foreign aid	Country	Remittances	Country	FDI
Liberia	48.4	Lesotho	32.9	Liberia	29.4
Afghanistan	37.4	Tajikistan	31.6	Hong Kong	28.3
Burundi	26.4	Moldova	26.0	Netherlands	25.2
West Bank and Gaza	25.8	Haiti	21.6	Ireland	20.1
Sierra Leone	20.8	Lebanon	19.9	Singapore	18.3
Eritrea	19.2	Nepal	18.2	Azerbaijan	17.5
Somalia	19.6	Jordan	17.7	Mozambique	14.0
Mozambique	19.1	Kosovo	17.5	Congo, Rep.	13.5
Rwanda	18.4	Kyrgyz Rep.	17.4	Mongolia	12.5
Malawi	16.8	El Salvador	16.1	Belgium	12.0

Notes: Calculations based on data from the World Development Indicators.

instances, these countries tend to be quite poor and/or have experienced civil conflict in the recent past (e.g., Liberia, Afghanistan, Rwanda).

In contrast, remittance recipients exhibit greater geographic variation. Since 2000, the top ten recipients of remittances lie in Africa (Lesotho), Eastern and Central Europe (Moldova, Kosovo), Central and Southeast Asia (Tajikistan, Kyrgyz Republic, Nepal), the Middle East (Lebanon, Jordan), the Caribbean (Haiti), and Central America (El Salvador). For some countries, remittances comprise more than a quarter of the country's national income (e.g., Lesotho, Moldova, and Tajikistan). Like aid recipients, the top remittance recipients tend to be poor and have often experienced political upheaval in the recent past, including episodes of political violence (e.g., Lebanon, Nepal) and emigration after gaining independence following the demise of the Soviet Union in 1990 (e.g., Moldova, Tajikistan, Kyrgyz Republic).

Whereas aid and remittance recipients tend to be low and low-middle income countries, the leading destinations for FDI tend to be in emerging markets in East Asia (e.g., Hong Kong is a gateway for investment into China), advanced economies in Western Europe with investor-friendly laws (e.g., Ireland, Belgium), and commodity- and resource-rich countries in Africa and Central Asia. The economic and political characteristics of these "FDI hotspots" differ. For example, while Western European countries are stable democracies, recipients in East Asia tend to be nondemocratic. The commodity-rich economies tend to be nondemocratic (e.g., Azerbaijan), relatively poor (e.g., Mongolia), and conflict prone (e.g., Liberia).

TABLE 2.2 *Composition of international capital, 1970–2015*

	Percentage share of total international capital		
	Aid	Remittances	FDI
1970–1979	54.5	21.6	23.9
1980–1989	52.5	27.7	19.8
1990–1999	42.8	26.4	30.8
2000–2009	31.1	29.2	39.7
2010–2015	27.7	32.4	39.9
1970–2015	39.7	28	32.3

Notes: Shares calculated by the author using data from the World Development Indicators.

Composition. The composition of capital flows also exhibits significant variation across countries and time. Table 2.2 reports the average share of each capital flow (relative to total capital flows) across all recipient countries since 1970. Over the entire sample period (1970–2015), foreign aid has comprised nearly 4 out of every 10 dollars of total capital inflows. The remaining 6 dollars are split roughly between FDI and remittances (with respective shares of 32.3 and 28 percent). These shares, however, have changed over time. Prior to 1990, foreign aid comprised slightly more than a half of foreign capital. Since then the aid share has nearly halved to about 27.7 percent. This is not due to an overall decline in aggregate foreign aid but to the tremendous growth of remittances and FDI (as depicted in Figure 2.1). Correspondingly, the remittance share has steadily increased from around 22 percent in the 1970s to 32 percent since 2010. The FDI share has exhibited even greater growth. It jumped from 24 percent in the 1970s to 40 percent since the 2000s. For many countries, remittances and FDI now comprise a greater share of their international capital than does foreign aid.

Authoritarian Politics

Having described the main independent variables, I now turn to the book's main dependent variable: authoritarian politics. To do so requires a measure to differentiate democracies and autocracies on their institutional characteristics, as well as one that captures the main objective of *all* governments (leaders): political survival. Fortunately, scholars have compiled data sets to measure these variables for a large number of countries, and over extended periods of time.

Institutional Variation. One prominent measure of democracy – in particular its *institutional* variation – is the POLITY score from the POLITY IV

2.1 Measurement

project.[19] The POLITY IV project is comprehensive in its scope and detail, and is largely consistent with some recent data sets that also code for political regimes.[20] The POLITY score comprises four subindices that measure various dimensions of political authority. They include the *competitiveness* of executive recruitment, the *openness* of executive recruitment, *constraints* on the chief executive, and the *competitiveness of political participation*.[21] In general, countries that have political institutions where leaders face fewer constraints on their authority, more restrictive barriers to entry and competition from rival leaders, and less political participation (e.g., elections) from the general population are classified as being less democratic.

The POLITY – in particular, the revised POLITY2 – score is ordinal, ranging on an integral −10 to +10 scale. Higher index values correspond to a greater quality of democratic institutions, and scholars tend to classify a country as being democratic if its POLITY2 score equals or exceeds +6. The data span 160 countries. For some countries, the data extend back to 1800 (e.g., Western European states and independent countries in North and South America), while for most states, the data start when the country gained independence (e.g., 1960 for many African countries, 1990 for former states and satellites of the Soviet Union). Since reliable cross-national data on international capital are available from 1970s onwards, I primarily analyze data from POLITY IV from 1970s onwards.[22] (In the subsequent empirical chapters, I explain these measures of institutional characteristics in greater detail.)

---

[19] As Marshall et al. (2015) state, the main objective of the project is to code "the authority characteristics of states in the world system for the purposes of comparative, quantitative analysis" (iii). Series from POLITY IV has been used extensively in empirical research in economics and political science. For instance, the data form the basis of some of the cross-national correlations underlying Acemoglu and Robinson's (2006) work on the economic origins of dictatorship and democracy. Bueno de Mesquita et al. (2003) employ data from POLITY IV to construct measures pertinent to testing their "selectorate theory." Alternate data sets are also available, including Cheibub et al. (2010), Boix et al. (2013), Coppedge et al. (2014), and Geddes et al. (2014).

[20] Some of these new data sources include Boix et al. (2013) and Geddes et al. (2014). Unlike these new data sets, the data from POLITY IV offer more variation in measuring regime type. Boix et al. (2013), for instance, code each country–year observation as either a democracy or an autocracy. Geddes et al. (2014) identify the different types of autocracy (e.g., military, monarchy) but do not code country–year observations for countries that are democratic. I use data from POLITY IV because they measure a country's regime characteristics across a wide range of institutional features (e.g., constraints on executive authority, restrictions on political participation). For instance, I use data from POLITY IV to identify strong and weak autocracies, as well as consolidated democracies.

[21] See Marshall et al. (2015) for a more detailed discussion on these subindices.

[22] In addition to evaluating the relationship between composite POLITY2 score and international capital flows, subsequent analyses in this book will examine the relationship between the various subindices and international capital. For instance, Chapter 5 evaluates the causal impact of remittances on executive constraints in autocracies.

TABLE 2.3 *Political institutions and the tenure of leaders since 1960*

Regime type	POLITY2 range	No. of leaders	Average tenure of leaders (days)
Strong autocracy	−10 to −6	220	4,040.2
Intermediate autocracy	−5 to 0	179	2,643.9
Weak autocracy	0 to +6	127	1,704.2
Democracy	6 to 10	490	1,483.2

Notes: Authors' calculations using data from POLITY IV and ARCHIGOS for leaders since 1960.

Political Survival. Having reliable measures of political institutions matters because they provide constraints on what governments (leaders) can effectively do.[23] Of course, whatever their particular objectives may be, leaders must first survive in political office. One such measure of political survival is the *number of days* a leader lasts in political office from ARCHIGOS.[24] For some leaders, especially those in democracies, their tenure tends to be shorter because of regularized elections and term limits. In contrast, leaders in autocracies face fewer institutional checks on their power and tend to last longer in political office. This is evident in Table 2.3, which lists the average duration of leaders since 1960 under varying "intensities" of authoritarian political institutions. In strong autocracies, leaders last on average 4,040 days (approximately 11 years). As the intensity of authoritarian institutions wanes (i.e., a higher POLITY2 score), so do the average durations of leaders. In the weakest forms of autocracy, the typical dictator lasts about 4.7 years in office. This is slightly longer than the typical democrat. In democracies, leaders last about 4 years in office, on average. On balance, the pattern in Table 2.3 shows that leaders in more authoritarian polities last longer in office.

2.2 INTERNATIONAL CAPITAL AND AUTHORITARIAN SURVIVAL

Patterns in the Raw Data

The notion that autocrats can leverage international capital to their political advantage is evident in the *raw data*. As a preliminary evaluation of this conjecture, I classify governments by their institutional features in three categories: strong autocracies, weak autocracies, and democracies.[25] I then

---

[23] North (1991); Bueno de Mesquita et al. (2003); Acemoglu and Robinson (2006).
[24] Goemans et al. (2009).
[25] Strong autocracies are countries with average POLITY scores less than −6 since 1970. Weak autocracies are countries with average POLITY scores between −6 and +5 since 1970. Democracies are countries with average POLITY scores greater than +6 since 1970. These

2.2 International Capital and Authoritarian Survival

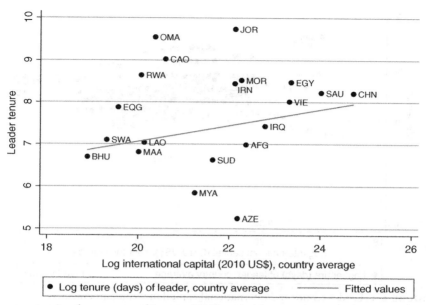

FIGURE 2.2 Leader tenure and international capital in strong autocracies.
Notes: Leader tenure and international capital is the country average since 1970. Sample of strong autocracies are countries with average POLITY2 scores less than −6.

examine the correlation between international capital and leadership duration across the three categories. (I exploit the *full range* of institutional variation and control for confounding factors in the next subsection.)

Figures 2.2, 2.3, and 2.4 illustrate these relationships. Each figure plots the bivariate association between leader tenure and the country's receipt of international capital across samples of strong autocracies (Figure 2.2), weak autocracies (Figure 2.3), and democracies (Figure 2.4). In each figure, the data are the country averages of leader tenure (log number of days) and international capital since 1970. For each country observation, international capital is the sum of its inflows of foreign aid, remittances, and FDI (in log units).[26]

For the sample of strong autocracies in Figure 2.2, the positive association between international capital and leader tenure suggests that a greater inflow of capital can permit leaders in strong autocracies to last longer in political

---

cutoffs (ranges) correspond to those used in existing scholarship such as consolidated autocracies, anoncracies, and consolidated democracies.

[26] The samples in Figures 2.2, 2.3, and 2.4 include countries that receive *all* three types of international capital. For example, the United States is not included in the analysis, as it does not receive any foreign aid.

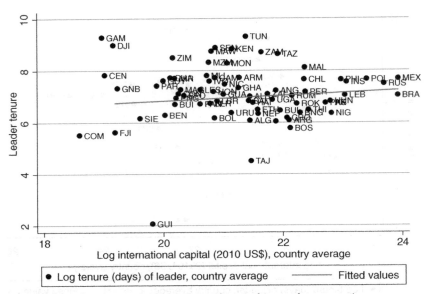

FIGURE 2.3 Leader tenure and international capital in weak autocracies.
Notes: Leader tenure and international capital is the country average since 1970. Sample of weak autocracies are countries with average POLITY2 scores between −6 and +5.

office.[27] The bivariate correlation is 0.34 and statistically significant (p-value = 0.0006). This positive association exists across a wide range of strong autocracies, including single-party Communist rulers (e.g., China, Vietnam), monarchies (e.g., Saudi Arabia, Jordan), military dictatorships (e.g., Egypt), and theocracies (e.g., Iran).

The positive association between leader tenure and international capital attenuates in countries with less entrenched authoritarian political institutions.[28] For a sample of weak autocracies, Figure 2.3 shows a *less* positive correlation between average leader tenure and capital inflows (ρ = 0.055). This weaker association reflects the larger heterogeneity in political institutional designs (e.g., military regimes, single-party rule, "hybrid electoral autocracies") as well as transitions from autocracy to democracy (and vice versa). For example, Brazil's military dictatorship in the 1970s and early 1980s actively courted FDI as

---

[27] Governments in strong autocracies enjoy low constraints on executive authority and impose restrictions on political participation from the general population.

[28] I follow convention and use POLITY scores above +5 as a cutoff to separate democracies from autocracies. This is for illustrative purposes only. Indeed, as Vreeland (2008) documents, these "cutoffs" can be problematic in empirical analysis. In the regression analysis to follow I circumvent this concern by using the full range of the POLITY index to evaluate the statistical relationship between international capital and political survival. Thus, the findings in this chapter are not contingent on any particular cutoff separating democracies from autocracies.

2.2 International Capital and Authoritarian Survival

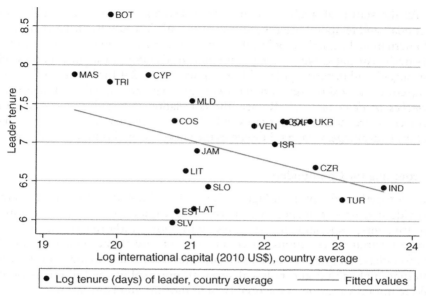

FIGURE 2.4 Leader tenure and international capital in democracies.
Notes: Leader tenure and international capital is the country average since 1970. The sample of democracies comprises countries with average POLITY2 scores equal to or greater than +6.

a strategy to promote economic development and co-opt the support of business elites.[29] While Brazil attained a POLITY2 score of +7 in 1985 following its successful transition to democratic governance, its average POLITY2 score since 1970 is +3.6, falling within the range of a weak autocracy. In other cases, some weak autocracies have cycled between democracy and autocracy. For instance, Bangladesh has received large amounts of foreign aid and remittance income and has fluctuated back and forth between democratic and autocratic rule since its independence.[30] Between 1971 and 2015, its average POLITY2 score is 0.80, with a maximal score of +8 in the early 1970s and a low of –7 in the mid-1970s and early 1980s.

---

[29] Evans (1979). Brazil's military dictatorship had an average POLITY2 score of –5.1 over its reign.
[30] After gaining its independence in 1971, Bangladesh emerged as a democracy with a POLITY2 score of +8. By the mid-1970s, however, the country had slid to authoritarian governance, with POLITY2 scores fluctuating between –4 and –7 from 1974 until 1989. In the 1990s, the country transitioned back to democratic rule (meeting the minimum threshold with a POLITY2 score of +6). Following a disputed election, a military "caretaker" government resumed power for two years (2008–2009) before relinquishing power to weakly and progressively less democratic government. Since 2009, the POLITY2 score has progressively declined from +5 in 2009 to +1 in 2015.

As the statistical analysis in the next subsection shows, the spatial (cross-sectional) and temporal heterogeneity in autocratic institutions mediates the effect of international capital on political survival. In particular, international capital is positively associated with a longer duration of dictatorial rule in countries with "stronger" authoritarian political institutions. Figure 2.4 further strengthens this conjecture: the positive association between international capital and leader tenure *vanishes in democracies*.[31] Rather, the correlation is negative, suggesting that democrats are not able to successfully leverage international capital to their political advantage. Chapter 3 discusses the reasons why this may be the case.

Accounting for Confounders

Taken together, the plots in Figures 2.2, 2.3, and 2.4 are informative, as they provide suggestive evidence that international capital can extend the tenure of governments, especially in countries with *more* authoritarian institutions. These bivariate correlations, of course, do not directly account for potential confounders, do not provide confidence intervals for any association between international capital and political survival, nor exploit the full cross-country and temporal variation in the quality of authoritarian institutions.

One strategy to improve the statistical precision of the relationships in these figures is to estimate a "survival model" that regresses a measure of leader survival on a country's receipts of international capital, a *continuous* measure of authoritarian institutions and their *interaction*, and a vector of potential confounders (e.g., economic growth, income).[32] In this specification, the inclusion of an interaction term captures the differential impact of international capital on leader survival as the country's institutions become *more* authoritarian. With this in mind, I estimate variants of the following econometric model:

$$EXIT_{it} = \alpha + \beta^* ICAP_{it} + \theta^* AUTOCRACY_{it} + \kappa^* (ICAP_{it} \times AUTOCRACY_{it}) + \lambda^* X_{it} + \tau_{it} + \varepsilon_{it} \quad (2.1)$$

---

[31] Figure 2.4 does not include many advanced economies that are democracies because they do not receive all three types of international capital. For instance, the figure omits the United States and Japan because neither country receives foreign aid.

[32] One strategy to model political survival is event history models (Box-Steffensmeier and Jones 1997). While there are several modeling approaches, a relatively parsimonious one is a "linearized hazard model" that regresses a measure of government "failure" on a set of parameters the analyst theorizes to affect the termination of a government's tenure (e.g., economic growth, institutional characteristics) and measures of "duration dependency." Ahmed (2012) discusses how a well-specified linearized hazard model (e.g., it includes measures of duration dependency) is an analog to other survival modeling approaches, such as Cox Proportional. Moreover, the flexibility of a linearized hazard model allows for estimation with fixed effects. Consequently, these models can account for various forms of unobservable confounders, such as time effects and unit-specific characteristics.

2.2 International Capital and Authoritarian Survival

where $EXIT_{it}$ is a binary variable equal to 1 if a leader exits political office in country i in year t, and 0 if the leader remains in office. Thus, a negative coefficient in equation (2.1) implies that a regressor *lowers* the likelihood of leader termination.

In equation (2.1) the key independent variables are each country's receipts of international capital ($ICAP_{it}$), an increasing measure of each country's prevailing "quality" of *autocratic* institutions ($AUTOCRACY_{it}$), and their interaction ($ICAP_{it} \times AUTOCRACY_{it}$).[33] This interaction term is appropriate to evaluate whether international capital has a differential effect on political survival in countries with more authoritarian political institutions. In particular, a *negative* coefficient on κ implies that higher amounts of international capital *lower* a leader's likelihood of exiting political office in countries with more authoritarian political institutions.

The specification includes a vector of time-varying country-level covariates that can affect a leader's prospect of political survival and inflows of international capital. In the baseline model, these confounders include a country's level of economic development (measured with its per capita gross domestic product [GDP]), growth rate, and population size. These data are from the World Bank's *World Development Indicators*. Following Carter and Signorino (2010), I account for how a leader's survival up to year t affects his or her future tenure (i.e., "duration dependency") with a polynomial (τ_{it}) of leader duration time (days in office). To account for potential serial correlation within country, I conservatively cluster the standard errors at the country level.

Table 2.4 reports the effect of international capital and political institutions corresponding to equation (2.1). Each column in Table 2.4 controls for confounding variables and a polynomial of leader duration terms. Column 1 shows that in isolation neither international capital nor autocratic institutions exhibit statistically significant effects on leader exit. However, their *interaction* exhibits a negative and statistically significant effect (coefficient = –0.001). This negative coefficient implies that, *on average*, higher amounts of international capital lower the probability of leader exit in autocracies. This coefficient estimate, however, masks how international capital affects leader exit across the *full* range of autocracy. To address this concern, methodologists recommend *graphing the entire range* of the interaction term to transparently illustrate the marginal effects.[34]

Figures 2.5 and 2.6 follow this recommendation by graphing the marginal effect – and the associated confidence intervals – of a unit change in international capital on leader exit as a country's institutions become *less*

---

[33] $AUTOCRACY_{it}$ rescales the POLITY index so a higher value implies greater autocracy. Specifically, $AUTOCRACY = POLITY2 - 21$. This means $AUTOCRACY$ ranges from –10 to +10, where a value of +10 implies the most autocratic countries.

[34] Brambor et al. (2006).

TABLE 2.4 *Marginal effect of international capital and political institutions on leader exit*

Dependent variable	Leader exit	
	(1)	(2)
Log international capital (2010 US$)	−0.007	−0.012
	(0.004)	(0.007)*
Autocracy	0.013	0.015
	(0.008)	(0.010)
Log international capital × Autocracy	−0.001	−0.001
	(0.0004)**	(0.0005)**
Time-varying country characteristics	Yes	Yes
Leader duration polynomial	Yes	Yes
Country fixed effects	No	Yes
Year fixed effects	No	Yes

Notes: $N = 2{,}255$. Robust standard errors, clustered by country reported in parentheses. *, ** = statistically significant at 10% and 5% respectively. Both specifications control for log GDP per capita (2010 US$), GDP per capita growth (% annual), log population, and a polynomial of leader duration (t and t^2). These coefficients and a constant are not reported. The specification in column 2 includes country and year fixed effects. These coefficients are not reported.

democratic.[35] Figure 2.5 graphs the marginal effect associated with the interaction of autocracy and international capital from column 1 in Table 2.4. While the marginal effect is negative over the entire range of *AUTOCRACY*, it is statistically significant only for "intermediate" and "strong" autocracies. Specifically, the confidence interval does *not* contain zero *only* when *AUTOCRACY* is *non*-negative. When *AUTOCRACY* is negative, the effect of international capital on leader exit is statistically indistinguishable from zero. Stated plainly, international capital lowers the likelihood of leader termination in strong and intermediate autocracies only. In *democracies*, international capital does *not* have a robust effect on leader exit.

Skeptics may worry that the omission of country-specific and temporal unobservables biases the findings. This may include time-invariant country

---

[35] Along the x-axis, the measure of autocratic institutions is a country's "inverted" (rescaled) POLITY2 score. Specifically, a higher value on this autocracy scale *(AUTOCRACY)* implies institutions with highly authoritarian institutional characteristics, such as extremely low constraints on executive authority, restrictions on rival leaders to compete for political power, and limitations on political participation from the broader population. For instance, "strong autocracies" are those with values of *AUTOCRACY* between +6 and +10, which corresponds to a POLITY2 score between −6 and −10.

2.2 International Capital and Authoritarian Survival

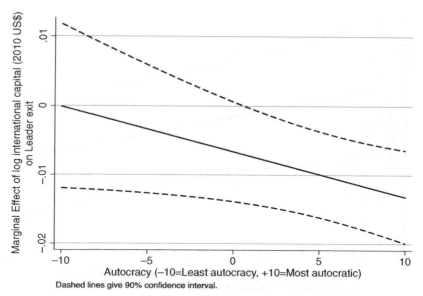

FIGURE 2.5 Marginal effect of international capital on leader exit at varying levels of autocracy.
Notes: Marginal effect corresponding to the interaction term in Table 2.4, column 1. Estimation with robust standard errors, clustered at the country level.

characteristics that can affect political survival such as the long-term consequences associated with the legacies of colonial institutions, climate, and geography.[36] One strategy to account for these unobservables is to estimate fixed-effects models with unit-specific (i.e., country) and time (i.e., year)-fixed effects. The inclusion of country fixed effects implies the regression coefficients evaluate the within-country effect of the covariates on leader termination. The inclusion of year fixed effects accounts for "temporal shocks" that can affect all countries, such as the oil price shocks of the 1970s and the end of the Cold War.

Column 2 in Table 2.4 reports the effects of international capital and political institutions from this "two-way fixed effects" model. In this specification, the interaction of international capital with autocracy is negative. This coefficient estimate, of course, is the *average* effect. Figure 2.6 decomposes this effect across the full range of autocracy. The figure reveals that international capital exhibits a robust negative effect on leader exit in countries with increasingly autocratic institutions. The implication is the same as in Figure 2.5: *international capital lowers the likelihood of leader exit in*

---

[36] Acemoglu et al. (2001).

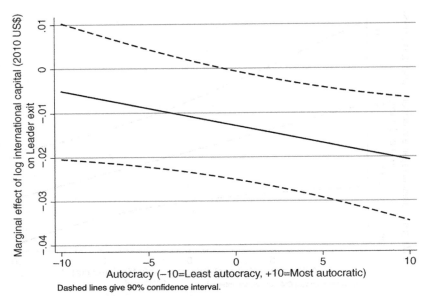

FIGURE 2.6 Marginal effect of international capital on leader exit at varying levels of autocracy, with country and year fixed effects.
Notes: Marginal effect corresponding to the interaction term in Table 2.4, column 2. Estimation with robust standard errors, clustered at the country level.

countries with intermediate and strong autocratic institutions. International capital does not affect leader exit in democracies.

On balance, Figures 2.5 and 2.6 provide greater statistical precision that international capital lowers the probability of leader termination in countries with more entrenched authoritarian institutions. In subsequent chapters, I will "unpack" this result for each separate capital flow.

2.3 NEXT STEPS

Globally, cross-border flows of capital have grown substantially since the early 1970s. These capital flows can take many forms: some are transfers between governments (foreign aid), households (remittances), and firms (foreign direct investment). Moreover, there is substantial heterogeneity in the "types" of recipients of each capital flow. For example, examining the wealth of recipient countries reveals that countries with low per capita income are the main recipients of foreign aid and remittance income, while higher income countries are hot spots for greater FDI.

Given the heterogeneity and sheer magnitude (e.g., exceeding 15 percent of national income), it is plausible that international capital can affect political

2.3 Next Steps

outcomes. Figures 2.2 to 2.4 explore one possible relationship in the raw data: the correlation between capital inflows and leader tenure in countries with democratic and autocratic institutions. These correlations, in corroboration with more sophisticated statistical techniques, provide preliminary evidence that higher amounts of international capital can extend the tenure of political leaders, especially in countries with more authoritarian political institutions.

The relationships in Figures 2.2 to 2.6 raise two important questions. The first question is theoretical: why should there be a relationship between foreign aid, remittances, and FDI and a government's political fortunes, especially in nondemocracies? And relatedly, can these capital flows also affect a country's underlying political institutions? In particular, what are the *channels* through which foreign aid, remittances, and FDI affect authoritarian politics? After all, these capital flows are received by different actors in an economy and are therefore likely to affect a government's political fortunes in different ways.

The second question is empirical: do foreign aid, remittances, and FDI have a *causal impact* on political outcomes? While suggestive, the relationships in Figures 2.2 to 2.6 are merely correlational; they are not causal. They do not mitigate the biases introduced by the endogeneity of international capital inflows and politics. For example, multinational firms prefer to invest in countries where property rights are more secure. These countries tend to be democratic.

Chapter 3 addresses the first question. The chapter emphasizes a distinct channel through which each capital flow affects authoritarian politics. Chapters 4 to 6 address the second question. As the theory identifies a distinct channel for each capital flow, I devote a separate chapter for each capital flow.

3

Foreign Rents and Rule

The trends in Chapter 2 provide preliminary evidence of a positive association between international capital inflows and leader survival in less democratic polities.[1] But why should inflows of foreign aid, remittances, and foreign direct investment (FDI) affect the fate of leaders, and do so *more* in nondemocratic regimes? After all, these capital inflows are received by various actors within an economy and do not all necessarily accrue directly to governments.[2] This chapter develops a theory that explains how governments can use capital inflows to fund strategies of political survival. In particular, international capital can embody nontax properties that enable leaders to finance *repression* and accumulate *loyalty*.

Each capital inflow can augment a government's revenue base through a distinct channel. A government can siphon some fraction of foreign aid into its revenue base, constituting an *income effect*. On the other hand, largely untaxed remittance income (which accrues to households) can allow a government to reduce its provision of welfare goods and divert those expenditures to policies that fund repression and accumulate loyalty, representing a *substitution effect*. By contrast, FDI can increase a government's revenue in two ways: through higher wages that a government can tax and by creating opportunities for government rent extraction. The former represents a *rentier effect*. As this chapter argues, the incentive for a government to employ these channels is *magnified* in less democratic countries. Accordingly, dictatorial rule can be magnified by international capital.[3]

---

[1] Throughout this chapter (and the rest of the book), I use the terms "governments" and "leaders" interchangeably.
[2] While governments are the main recipients of foreign aid, remittance income enriches households, and FDI can improve the performance of domestic firms.
[3] The interaction between the government, firms, and households can be viewed as a sequential process (or "game") involving two stages. In the first stage, firms and households optimize their use of FDI and remittances, respectively. In the second stage, the government "observes" these decisions by firms and households and responds optimally in deciding its own spending decisions.

This chapter develops this theoretical proposition from the ground up. I start by identifying two fundamental political problems faced by dictators – the problems of authoritarian control and power-sharing – and the primacy of revenue to help them mitigate these challenges.[4] In this first section, a distinction is drawn between tax and nontax revenues and their relevance for understanding various political outcomes (e.g., political stability, democratic accountability). Section 3.2 then discusses two broad types of strategies governments pursue to preserve their incumbency: repression and loyalty. This section describes various indicators of repression and loyalty. measures that will be used in the subsequent empirical chapters to test the theory. Building on these foundations of political survival, Section 3.3 describes how differences in political institutions (e.g., regime type) affect how governments allocate their revenues on repression and patronage. The section argues that governments in less democratic countries are more predisposed to allocate their spending on repression and patronage. And in doing so, a dictator can better overcome the problems of authoritarian control and authoritarian power-sharing discussed in Section 3.1.

Section 3.4 integrates international capital into this framework. It argues that foreign aid, remittances, and FDI can finance repression and patronage via three distinct channels: an income effect from foreign aid, a substitution effect associated with remittances, and a rentier effect stemming from FDI.[5] None of these channels involve directly taxing the domestic population. Thus, foreign aid, remittances, and FDI can represent sources of nontax income, which in turn can help leaders preserve their rule. In Section 3.5, I discuss potential theoretical and empirical challenges and how the subsequent empirical chapters will address these concerns. Section 3.6 summarizes the chapter.

3.1 FOUNDATIONS OF POLITICAL SURVIVAL

The Dual Problems of Dictatorial Rule

All leaders strive to stay in power, whether in a democratic or nondemocratic political environment. However, dictators are particularly sensitive to two problems: the problems of authoritarian control and authoritarian power-sharing.[6] Each problem relates to a distinct group in a polity: those who rule and those who are ruled. All dictators face threats from the masses, and the challenge of balancing against the majority excluded from power constitutes the problem of authoritarian control. In practice, of course, dictators rarely control enough resources by themselves to thwart these challenges and therefore rule

---

[4] This discussion draws heavily on Levi (1988) and Svolik (2012).
[5] These insights are formalized in a parsimonious game-theoretic model in the chapter's appendix.
[6] The discussion in this subsection builds extensively on Svolik's (2012) rich theoretical exposition on the politics of authoritarian rule.

with a number of allies, such as traditional elites, prominent party members, or military officers. This can generate a second, separate political conflict when dictators face challenges from those with whom they share power. This is the problem of authoritarian power-sharing.

A dictator who fails to address these problems heightens the prospect of losing power, typically through elite defection (e.g., coups) and less frequently through revolution.[7] A dictator can employ a variety of strategies to tackle these problems. These strategies – as I discuss in greater depth shortly – can be classified as *repression* and *loyalty*. While the optimal mix of these strategies will vary across countries, their very utilization presupposes that a government has the means to finance these strategies.[8] Thus, revenue is a first-order concern for rule.

The Primacy of Revenue

Government action requires economic resources. Taxation has been the most common way to acquire them. For Cicero, taxes comprised "the sinews of the state" upon which the Roman Empire relied. For social scientists, revenue and taxation are central for state capacity and development.[9] Rudolf Goldscheid, the founder of fiscal sociology, avers that "the budget is the skeleton of the state stripped of all misleading ideologies" (quoted in Schumpeter [1918] 1954, 6). Echoing a similar sentiment, Schumpeter argues, "the fiscal history of a people is above all an essential part of its history" ([1918] 1954, 6–7). Indeed, for those at the helm of government, revenue underlies rule: "the greater the revenue of the state, the more possible it is to extend rule."[10]

With sufficient revenue, a ruler enhances his or her ability to develop the institutions of the state (e.g., military, civil service), incorporate more individuals within the domain of those institutions, and expand the supply and variety of public goods. In pursuit of this objective, all rulers are "predatory" in that they strive to extract as much revenue as possible to achieve their social and personal ends.[11] Revenue can be outright stolen by a ruler, used to finance the aggrandizement of his or her power, provide collective goods, and/or achieve other public policies.[12]

---

[7] Svolik (2012, 5) documents that over the period 1946–2008, nearly seven of ten dictators exited office via coups. In contrast, only 11 percent of dictators exited office due to popular uprisings. Assassinations, foreign interventions, and transitions to democracy are less frequent sources of leader exit in nondemocracies.

[8] The variation in policies stems in large part from differences in political institutions. I will describe how so shortly.

[9] For more recent treatments on the centrality of taxation in state capacity and governance, see, for example, Levi (1988), Tilly (1992), Besley and Persson (2010), Profeta and Scabrosetti (2010), and Morrison (2015).

[10] Levi (1988), 2.   [11] Ibid., 3.

[12] Levi is clear that a ruler can use revenues to benefit him- or herself and/or his or her subjects. Rulers "may use the funds to line their own pockets or to promote their personal power. They may use the funds to support social or personal ends. They may have ideological ends they wish

3.1 Foundations of Political Survival

Rulers face constraints, however. Maximizing state revenue involves the costs of extraction or a trade-off between extraction costs (e.g., deadweight loss associated with taxation) and aggregate income (e.g., growth). This requires minimizing popular protests and reducing the transaction costs associated with taxation. Given these concerns, rulers may have an incentive to reduce the tax burden of their citizens and derive revenue from alternate sources.

Broadly speaking, a government can collect revenue through two channels: taxation and nontax sources. The former may entail the extraction of income from individuals (e.g., wages, salaries), firms, land and capital owners, and associated economic transactions via consumption (e.g., value-added taxes) and international trade (e.g., tariffs), among others. The latter may entail revenue derived through the state's production of natural resources (e.g., oil, minerals, forests, diamonds) and external transfers (e.g., economic and military assistance). While there is a voluminous literature on the political economy of taxation, scholars are increasingly probing how nontax government revenue influences politics.[13]

Nontax revenue can include revenues from oil because it can be derived through the operation of state-owned oil enterprises rather than through taxes.[14] Nontax revenue also includes foreign aid; borrowing; intergovernmental grants at the subnational level; and all other nontax government income from other state-owned enterprises, fines, and so forth.[15] Furthermore, as this chapter will argue, international capital that does not flow directly to a government can constitute nontax revenue – through the appropriate channel – and exhibit similar effects on politics.

The availability of nontax revenue is especially important in many developing countries where governments often lack the capacity to effectively tax their citizens. Morrison (2015) calculates that nontax revenue accounts for an average of 27 percent of national government expenditures, with this figure higher if one includes nongovernmental international capital flows, such as inward FDI and remittances.[16] At the subnational level, nontax revenue often

---

to promote. They may be altruistic. Randomness of ends characterizes rational choice models in general, and rulers operate with a wide range of them. It is by means of the state and its revenues that rulers achieve their personal and social ends" (Levi 1988), 3.

[13] On the political economy of taxation, see Brennan and Buchanan (1980), Meltzer and Richards (1981), Levi (1988), Besley and Persson (2010), and Profeta and Scabrosetti (2010). I use the terms "nontax revenue" and "unearned income" interchangeably. Unearned income is a concept in economics that has different meanings and implications depending on the theoretical framework used. Political economists broadly typically view unearned income as nontax *government* revenue (e.g., Mahdavy 1970; Morrison 2015; Besley and Persson 2010).

[14] State-owned enterprises control 75 percent of the world's oil production and 90 percent of its reserves (Ivanhoe 2000; Tordo et al. 2011). Examples include Saudi Aramco, China National Petroleum Corporation, Nigerian National Petroleum Corporation, and Petroleos de Venezuela.

[15] Morrison (2015), 4.

[16] In some countries, nontax income (e.g., revenues from natural resources, state-owned enterprise, fines, foreign aid) accounts for an exceedingly large share of government expenditures. Based on

is the dominant source of local government revenues.[17] Calculations based on data from Rodden (2004) show that nontax revenue comprises a significant share of local government revenue in a variety of countries including the United States (61%), Mexico (91%), Argentina (82%), Nigeria (91%), Botswana (99%), India (66%), Philippines (94%), and Australia (67%).[18]

The Politics of Nontax Revenue

Extracting nontax and tax revenue has implications for political development. Tilly (1992), for instance, argues that modern European states emerged due to war: "war made the state and the state made war." To finance war, rulers had to raise revenue from their populations, which required bargaining between revenue-maximizing leaders and their citizens. For Tilly, the political bargains governments make hinge on the distribution of this income in exchange for staying in power, as opposed to relinquishing some influence over policy choice in exchange for taxes. With more warfare over time and thus a greater reliance on tax revenue, governments tended to become more accountable to their populations.

This relationship between nontax revenue and political accountability is borne out in the data. One important dimension of political accountability is a country's quality of political freedoms.[19] As an illustration of this, Figure 3.1 shows that governments that derive a greater share of their revenues from taxes extend greater political rights to their populations. The measure of political rights (on the x-axis) uses scores from Freedom House that hone in on such rights.[20] A similar pattern holds using a country's POLITY score: democracies derive a larger share of their revenue from taxes. In Figure 3.1, governments in the "most free" countries (e.g., Costa Rica, France, Japan, and the United States), on average, derive about 27 percent of revenues from taxation on income and profits.[21] For some governments, taxes contribute more than 50 percent of their revenue (e.g., the

---

data averaged over the period 1973–2001 (by Morrison 2015, figure 1.1), representative examples include Nicaragua (35%), Bolivia (42%), Republic of Congo (59%), Burundi (44%), Greece (32%), Romania (47%), Syria (50%), Iran (67%), Nepal (54%), Pakistan (38%), and Japan (28%).

[17] For instance, countries with a federal system rely on intergovernmental grants to finance local governments, and many of these local governments receive more in grants than they pay out in taxes to fund these transfers (Broadway 2008).

[18] These shares are reported in Morrison (2015), figure 1.3.

[19] On the importance of political rights (repression) and democracy, see Wintrobe (1998) and Davenport (2007a, 2007b).

[20] In the literature, scholars view this share as a broad, cross-national measure of "tax effort." The measure of political freedom is from Freedom House. The negative relationship in Figure 3.1 holds when controlling for a country's level of economic development, as well with alternate measures of political accountability/freedom such as Freedom House's measure of "civil liberties" and POLITY's measure of "political participation."

[21] This excludes consumption-based taxes, which also comprise a significant share of revenues in these countries.

3.1 Foundations of Political Survival

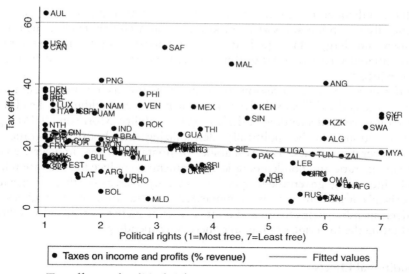

FIGURE 3.1 Tax effort and political rights.
Notes: The figure plots average revenues collected from taxation on income and profits (% revenue) and Freedom House's measure of political rights for each country.

United States, Canada, and Australia). In contrast, governments in the "least free" countries (e.g., Afghanistan, Belarus, Syria, Vietnam), on average, collect about 17 percent of their revenues from taxation on income and profits. This negative relationship between tax effort and political accountability underlies theories of the "rentier state" and the resource curse, which posits that governments that do not rely on taxation for their revenues tend to be more insulated from their populations, less accountable to them, and less democratic.[22]

The relationship between nontax income and democracy emerged from scholars of the Middle East investigating whether the prevalence of rents contributed to authoritarian governance in that region.[23] The central argument in the rentier state literature is that governments funded by external rents are freed from the need to raise taxes, which makes them less accountable to their citizens, and hence less likely to deploy these rents in ways that promote economic development. Focusing on the political ramifications, political economists have refined the theoretical channels through which rents can

---
[22] Mahdavy (1970); Beblawi (1987); Ross (2001, 2013); Robinson et al. (2006).
[23] Mahdavy (1970, 428) first postulated the contemporary conception of the rentier state as a state that receives substantial rents from foreign individuals, concerns, or governments. Beblawi (1987, 50) developed a more precise conceptualization of a rentier state. In such a state, (1) the rents are paid by foreign actors, (2) these rents accrue directly to the state, (3) only a few domestic actors are engaged in the generation of this rent (wealth), and (4) the majority are only involved in the distribution or utilization of the wealth.

foster nondemocratic governance, such as funding repressive capacity and financially co-opting support from key elites and constituencies, such as the military and clergy. Many qualitative and quantitative studies have empirically substantiated a negative relationship between unearned income and democratic governance.[24]

Of course, there is no reason to assume that nontax revenue should affect governance and stability exclusively in nondemocracies. Building on the idea that *greater* government spending underlies regime stability, Morrison (2015) argues that taxation can be politically destabilizing for both democratic and nondemocratic regimes. The availability of nontax revenue, however, enables both democratic and nondemocratic governments to reduce their taxation levels while increasing government spending, thereby increasing their longevity.[25] Morrison's argument can be summarized as: "taxation hurts regimes and leaders, spending helps them, and nontax revenue increases political stability by facilitating the lowering of taxes and increasing spending."[26]

Conditional Effects

Yet, while Morrison shows that nontax income can be a source of political stability and may not always be a "political curse" (at least in democracies), preexisting institutional quality often affects whether nontax income fosters economic growth and political stability. For instance, Tornell and Lane (1999) advance a model showing how a state with weak institutions, on receiving a positive shock to fiscal capacity (e.g., a resource boom, foreign aid windfall) may suffer from a "voracity effect" in which powerful groups compete for and squander the windfall and diminish economic growth. Indeed, such predation may actually lower economic development in weak institutional environments. Mehlum et al. (2006) show that where ex ante institutions are "grabber friendly" (e.g., more prone to corruption), nontax income tends to lower aggregate income. In contrast, where the preexisting institutional environment is "producer friendly" (and less prone to corruption), nontax income raises aggregate income. Focusing on the political ramifications, Robinson et al. (2006) formulate a parallel model demonstrating that when institutions are weak ex ante, increases in nontax income lead to excessive public employment and patronage (e.g., corruption). In these weak institutional settings, nontax income can finance additional forms of patronage goods such as the co-option of elites (e.g., the clergy, rival political leaders) and the military.[27] By directing

---

[24] See Ross (2013) for an overview.
[25] Morrison provides compelling cross-national evidence, coupled with a case study of subnational spending in Brazil, supporting this conjecture. Regimes that rely on greater amounts of nontax revenue exhibit more "durable" political institutions and lower leadership turnover.
[26] Morrison (2015), 3.
[27] Moreover, Liou and Musgrave (2014, 1586) argue that "resource income enables leaders to more effectively implement their winning coalitions pre-windfall preferences within the confines

3.2 STRATEGIES

additional resources to the latter, it can strengthen government's repressive capacity to mitigate evolutionary threats.[28]

Practicing Repression

To mitigate the prospects of political demise by domestic actors, governments can pursue two broad strategies: *repression* and *loyalty*.[29] Following Wintrobe (1998, 33–34), repression refers to "the restriction on the rights of citizens to criticize the government, restrictions on the freedom of the press, restrictions on the rights of opposition parties to campaign against the government, or as is common under totalitarian dictatorship, the outright prohibition of groups, associations, or political parties opposed to the government."[30] Essentially, *repression implies a lower quality of political rights*.

In contrast, political loyalty refers to the accumulation of trust (or allegiance) by a government from its citizens and/or relevant constituent subgroups, such as the military, traditional elites, and coethnics. One way to accumulate greater loyalty is through the distribution of targeted government resources to these subgroups. Another strategy is to raise the material well-being of the masses. For example, by pursuing policies that encourage broad-based economic growth, a government can reduce the incentives for rebellion from the masses.[31] (Loyalty is discussed in greater detail in the next section.)

---

of existing institutional constraints." A conditional resource curse may arise from other – but related – features of a country's domestic political economy, including income inequality (Dunning 2008), ownership structure (Luong and Weinthal 2010), electoral institutions (Mahdavi 2015), and economic integration (Kurtz and Brooks 2011).

[28] Bueno de Mesquita and Smith (2010a).

[29] Although foreign interventions can be a source of leader removal, they are relatively rare compared to domestic sources, such as elections, natural deaths of leaders, coups, and internal rebellion (Bueno de Mesquita et al. 2003; Svolik 2012). Nevertheless, governments can reduce the likelihood of removal from foreign intervention through a number of strategies. For example, this can include strengthening the domestic military, joining security alliances and defense pacts, and "buying protection" (e.g., tacit agreement with the Saud family in Saudi Arabia to provide reliable flows of oil to the USA/West in exchange for "protection" from domestic and international threats to the Saud monarchy). Lake (2009) provides a rich theoretical and empirical account of how "international hierarchy" can impact a country's domestic politics.

[30] Engaging in repression requires sufficient state capacity and revenues. As Wintrobe (1998, 34) observes, for repression to be effective, "these restrictions must be accompanied by the monitoring of the population and by sanctions for disobedience. The existence of a political police force and of extremely severe sanctions for expressing and especially for organizing opposition to the government (such as imprisonment, internment in mental hospitals, torture, and execution) represents the hallmark of dictatorships of all stripes.

[31] Stagnation or declines in personal income can heighten revolutionary episodes. See, for example, the historical evidence in Acemoglu and Robinson (2006).

In theory, governments can engage in various combinations of repression and loyalty. Given their institutional constraints, however, repression is not a particularly prudent strategy for governments in democracies. For example, democratic leaders face stricter controls on their power (e.g., legal statutes that limit executive powers, checks and balances from the legislature and courts), and the excessive suppression of political rights is antithetical to democratic principles. In contrast, repression can be a viable strategy of political survival in autocracies where such institutional constraints are weaker. In particular, as Wintrobe (1998) formalizes in his economic theory of dictatorship, increasing repression to sufficiently high levels enables a dictator to reduce the ability of his or her citizens to organize and/or co-opt other domestic actors (e.g., the military) that could help depose of the dictator.[32] In effect, high levels of repression can stall domestic rebellion in nondemocracies. Repression can be an effective instrument of authoritarian control.[33]

The prevalence of repression in autocracies (relative to democracies) is borne out in the data.[34] Although the level of political repression may be difficult to measure conceptually, a number of quantitative measures are available. Table 3.1 provides summary statistics comparing various reputable measures of political and human rights across autocracies and democracies. In Table 3.1 higher values of these measures correspond to greater repression. For example, Amnesty International – a reputable, international nongovernmental organization (NGO) – has evaluated countries on their human rights practices on an annual basis since 1976. Wood and Gibney (2010) have mapped these evaluations on an increasing 5-point scale of "political terror" (row 1).[35] Based on Wood and

---

[32] Raising repression, however, raises a potential challenge to the dictator. To pursue repressive tactics, "the dictator's security forces must themselves be given sufficient power, and their own uncertain reliability may, in the end, constitute the main threat to the ruler's regime. One obvious way to solve this problem is to overpay the security forces" (Wintrobe 1998, 34). These "overpayments" can be viewed as a form of patronage that a dictator provides to win the loyalty of his security forces. This form of patronage is discussed in the next section.

[33] Davenport (2007a, 2007b).

[34] The measures of repression between autocracies and democracies are relatively constant across time. For example, comparison of means and differences between autocracies and democracies for subsamples at decade frequencies (e.g., 1970, 1980, 1990, 2000, 2010) yields similar inferences as those in Table 3.1, which collapses all years.

[35] Wood and Gibney (2010) code five levels on the "political terror scale." Level 1 corresponds to countries under a secure rule of law, where people are not imprisoned for their views, torture is rare or exceptional, and political murders are extremely rare. Level 2 corresponds to a limited amount of nonviolent activity. However, few persons are affected, torture and beatings are exceptional, and political murder is rare. Level 3 corresponds to extensive political imprisonment, or a recent history of such imprisonment. Execution or other political murders and brutality may be common. Unlimited detention, with or without a trial, for political views is accepted. Level 4 corresponds to countries in which civil and political rights violations have expanded to large numbers of the population. Murders, disappearances, and torture are a common part of life. In spite of its generality, terror on this level affects those who interest themselves in politics or ideas. Level 5 corresponds to countries where terror has expanded to the

3.2 Strategies

TABLE 3.1 *Political repression in autocracies and democracies*

	Autocracy	Democracy	Difference	Range
	(1)	(2)	(3)	(4)
PTS – Amnesty International	3.056	2.253	0.802	[1, 5]
	(0.024)***	(0.018)***	(0.030)***	
PTS – US Department of State	2.889	1.975	0.914	[1, 5]
	(0.018)***	(0.020)***	(0.027)***	
Political rights	5.632	1.865	3.767	[1, 7]
	(0.021)***	(0.019)***	(0.029)***	
Civil liberties	5.287	2.222	3.065	[1, 7]
	(0.020)***	(0.021)***	(0.029)***	
Executive constraints	5.56	1.432	4.128	[1, 7]
	(0.019)***	(0.012)***	(0.026)***	

Notes: Standard errors reported in parentheses.
*** Significant at 1%. Higher values on the political terror scale (PTS), political rights, civil liberties, and executive constraints indices correspond to greater political repression. In column (1), autocracies are countries with a POLITY score less than +6. In column (2), democracies are countries with a POLITY score of +6 or greater. "PTS" refers to the political terror scale. "Political rights" and "civil liberties" are from Freedom House. "Executive constraints" is from POLITY IV.

Gibney's coding, the typical autocracy has extensive political imprisonment where execution or other political murders and brutality may be common (PTS = 3), while democracies have a limited amount of imprisonment for nonviolent political activity. The difference in means (= 0.8 point) is statistically significant, implying that autocracies are more repressive than democracies.

A similar relationship holds using an analogous measure of political terror based on annual reports by the US Department of State (row 2) and Freedom House's annual scoring of political rights and civil liberties (rows 3 and 4).[36]

whole population. The leaders of these societies place no limits on the means or thoroughness with which they pursue personal or ideological goals.

[36] The definitions/classifications of the US Department of State levels of political terror are identical to those of Amnesty International (Wood and Gibney 2010). The indices are highly correlated. The Freedom House measures are based on an annual survey of the state of global freedom as experienced by individuals. The survey measures freedom according to two broad categories: political rights and civil liberties. Political rights enable people to participate freely in the political process, including the right to vote freely for distinct alternatives in legitimate elections, compete for public office, join political parties and organizations, and elect representatives who have a decisive impact on public policies and are accountable to the electorate. Civil liberties allow for the freedoms of expression and belief, associational and organizational rights, rule of law, and personal autonomy without interference from the state. The political rights and civil liberties measures lie on a 7-point scale (1 to 7), with 1 representing the highest and 7 the lowest level of freedom.

The consistent patterns across these measures of political rights – and by different rating agencies (e.g., the US government, NGOs, and academic scholars) – demonstrates how the quality of these rights differ across countries with varying political *institutions* (i.e., democracies vs. autocracies).

In addition to focusing on the rights of individuals, political repression can also be evaluated through the *political constraints* faced by governments (leaders). These also vary by a country's institutions. Using the executive constraints index from POLITY IV, row 5 in Table 3.1 shows that leaders in autocracies face fewer constraints on their rule than democratic leaders, and these lower constraints on executive power are a major factor differentiating autocracies from democracies.[37]

Accumulating Loyalty

Governments can also remain in power by accumulating political loyalty from their supporters, while the failure to do so can lead to the government's removal from power via peaceful (e.g., elections) or violent (e.g., military coup, a loss in civil war) means. For instance, the notion of loyalty plays a central role in selectorate theory.[38] In this theory, two groups of people affect a leader's prospects of political survival. These groups are the selectorate (S) and a subset, the winning coalition (W).[39] To remain in power, leaders must maintain the loyalty of their winning coalition through the distribution of targeted benefits (private goods) and public goods. In autocracies where the winning coalition is small, the leader will provide private goods to satisfy the coalition. As the size of the winning coalition increases, such as in democracies, the leader will supply public goods to satisfy the larger coalition.

Leaders supply private goods to prevent defection of members in the winning coalition to a challenger. Whereas remaining loyal to the incumbent ensures the continuation of benefits from the leader, defection to a challenger entails some uncertainty about future benefits. As Bueno de Mesquita et al. (2003, 66) state explicitly: "The risk of exclusion from a challenger's long-term winning coalition drives loyalty to the current leader." The strength of this "loyalty norm" to the incumbent leader depends on the ratio W/S.[40] In particular, "the

---

[37] Gledistich and Ward (1997).   [38] Bueno de Mesquita et al. (2003), 65–68.

[39] The selectorate (S) includes every eligible person who has some say in choosing the leader (government). For example, in an American presidential election, S is the people who cast a vote. The winning coalition (W) is the subset of individuals from the selectorate whose support translates into a victory. For example, in an American presidential election, W is the group of voters who get a candidate to 270 Electoral College votes. In other countries, leaders may stay in power with the support of a smaller number of people, such as the clergy (e.g., Saudi Arabia, Iran) and senior figures in the security forces (e.g., military dictatorships) and business oligarchs (e.g., contemporary Russia).

[40] As Bueno de Mesquita et al. (2003, 67) note: "the risk of exclusion is central to the selectorate theory. In the context of our mathematical model, we derive the probability that a selector is

3.2 Strategies

smaller W/S is (and hence the larger risk of exclusion from a future coalition, 1 − W/S), the less inclined any member of the coalition is to put private benefits at risk by giving support to a political opponent of the incumbent."[41] As autocracies are characterized by a small winning coalition and a large selectorate, autocratic governments are especially incentivized to accumulate loyalty. In practice, leaders can employ a number of strategies to mitigate the likelihood of defection and better ensure their political survival including distribution of *patronage* and acquisition of *legitimacy*. These strategies – distributing patronage and accumulating legitimacy – are elaborated in the text that follows.

Patronage. In many instances, a government attempts to buy loyalty through the distribution of rents to key constituents.[42] This patronage helps mitigate the problem of authoritarian power-sharing. Patronage can take a variety of forms. For instance, a government can extend an outright transfer of wealth or power/right to extract wealth from other sources (e.g., bribery) to its key constituents, such as the military, religious and business elites, bureaucrats, and the general public. Such patronage could entail the allocation of subsidies, licenses, and contracts to certain individuals, ethnic and religious groups, industries, and business elites in exchange for their political support to the regime. The mix of these types of patronage will vary by country. In many African countries, for example, ethnicity drives political decisions where coethnics of the ruling government receive greater transfers and other "goodies" from the government relative to ethnic groups that are dissimilar from the government.

In many instances, patronage could embody employing regime supporters in the public sector. In his empirical analysis of government spending, Keefer (2007, 809) argues that "patronage employment is a well-known staple of clientelist governments" and uses the government wage as "a measure of government incentives to channel spending to targeted constituencies." Using such a measure, there is a strong empirical relationship between the size of the public sector and nondemocratic governance. Figure 3.2 shows that governments in less democratic regimes allocate a greater share of their expenditures to the compensation of government employees. This relationship also holds when excluding potential outliers, such as countries from the Middle East (e.g., Bahrain, Syria, Iran, Jordan), as well as netting out the effect of economic development (e.g., GDP per capita).

Legitimacy. Loyalty can also be *earned* by a government. In doing so, a government can earn the trust of key subgroups (e.g., business elites, military,

---

included in future coalitions to be precisely W/S; hence the risk of exclusion and the concomitant loss of future private goods is (1 − W/S)."

[41] Bueno de Mesquita et al. (2003), 68.

[42] This is especially the case in nondemocratic countries where doing so to a smaller "winning coalition" reduces their probability of defection (Bueno de Mesquita et al. 2003).

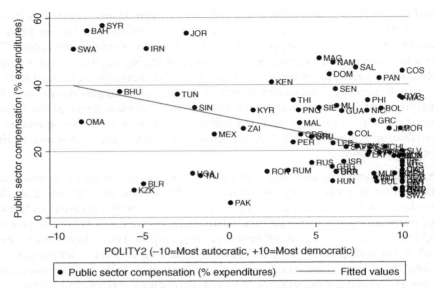

FIGURE 3.2 Public sector compensation and democratic governance.
Notes: The figure plots the average government expenditures on employee compensation (% expenditures) and POLITY2 score for each country.

coethnics) and/or the masses, who in turn are less likely to dispose of the incumbent.[43] This trust can be earned in a variety of ways, and may incorporate unique methods. For example, Emperor Augustus of ancient Rome committed adultery so "as to discover what his enemies were at by getting intimate with their wives or daughters."[44] In contemporary times, trust of the government can be garnered by its implementation of policies fostering economic growth and raising living standards – thereby making the ruling government more likely to stay in power.[45]

The importance of performance legitimacy is present in both democracies and autocracies. For instance, the capital-intensive and export-led strategies of numerous East Asian autocracies in the late twentieth century (e.g., Singapore, South Korea, Indonesia, Malaysia, China) are predicated on the

---

[43] Wintrobe (1998), 36.
[44] Wintrobe (1998), Augustus was one of the most skilled leaders at accumulating power the world has ever known, a ruler who (with Anthony) defeated Brutus and then subsequently Anthony (and Cleopatra), and founded the Roman Empire.
[45] The benefits of economic development do not have to be equitable, as elites may receive disproportionately more of the gains through, for example, preferential access to credit, licenses, and government contracts, or through capital gains. In some instances, the government may not want to pursue economic development at all and resort to repressing its population (e.g., Robert Mugabe's dictatorial rule in Zimbabwe).

belief that sustained economic development helps legitimize the regime's authoritarian practices and discourage calls for democratization. In democracies, theories of economic voting and corroborative cross-national evidence show that economic growth raises an incumbent government's likelihood of reelection.[46]

3.3 COMPARATIVE PUBLIC FINANCES

Public and Private Government Consumption

The capacity of nondemocratic governments to co-opt the relevant groups (e.g., business elites, the military, various ethnic and/or tribal groups) and/or repress the population is contingent on sufficient financial resources. Repression, for example, requires an extensive apparatus to monitor the population effectively and to levy sanctions for disobedience.[47] Patronage also requires sufficient financial resources. In autocracies, because these expenditures on repression and patronage are intended to benefit the leader (regime) and to support the regime's prospects of political survival, they can accordingly be viewed as a form of "private government consumption." This does not mean, however, that all autocratic governments allocate their resources to private consumption. Most autocracies spend varying amounts of government revenues on goods that benefit the broader population, such as expenditures on national security, public infrastructure, and some public services (e.g., health, education). These types of expenditures can be viewed as "public government consumption."

Existing theories predict that the relative mix or share of government expenditures allocated to public and private consumption tends to differ by the quality of democratic governance in a country.[48] For instance, Acemoglu and Robinson (2006) argue that a country democratizes as a credible commitment to future redistribution. By design, democratic governments will therefore spend a larger fraction of their revenue on the provision of public goods. Similarly, selectorate theory posits that given their political institutional constraints, governments in democracies where the winning coalition is large (relative to autocracies) will spend a greater share of revenues on the provision of welfare goods to benefit the masses. Doing so better ensures the political survival of democratic governments. This trade-off between patronage (private consumption) and welfare goods (public consumption) will enter the

---

[46] Lewis-Beck and Stegmaier (2000).

[47] A prerequisite for any government (leader) to engage in repression is an expansive set of executive powers. Without such authority, repression is a less optimal strategy (Wintrobe 1998). Wintrobe classifies these low repression states as "tinpots" whereby dictators rely more on strategies of loyalty than on repression to remain in power. The military dictatorships in Latin America during the 1970s and 1980s typify Wintrobe's conception of a tinpot regime.

[48] By definition, authoritarian regimes have a lower quality of democratic governance.

government's utility function.[49] Specifically, governments in autocracies will place greater weight on expenditures on private government consumption and correspondingly less weight on expenditures toward public government consumption.[50]

Moreover, there is variation in the quality of nondemocratic governance within autocratic regimes. For instance, totalitarian regimes (e.g., Nazi Germany, Stalin's Soviet Union) utilized high levels of repression and loyalty, whereas tinpot regimes (e.g., Somosa of Nicaragua, the Shah of Iran) utilized lower levels of repression and loyalty to maintain power.[51] Thus, the weight governments place on expenditures on private consumption will also vary across the different types of authoritarian regimes.

Examples

Welfare Goods. Empirically, there is a robust correlation between the quality of democracy and a government's expenditures on public consumption (welfare goods).[52] One such measure, for which data exist for a large number of countries, is government expenditure on transfers and subsidies to a large set of its domestic population. To capture this relationship, Figure 3.3 plots a country's average expenditures on government transfers and subsidies (as a share of its total expenditures) against its average POLITY score over the period 1990–2015, where a higher POLITY score implies a higher quality of democratic governance. There is a striking positive correlation between a country's POLITY score and its share of government expenditures allocated to public consumption. Governments in more democratic countries (e.g., Great Britain, Uruguay, and Poland) allocate a larger share of their government budgets to welfare spending than do less democratic countries such as Bhutan, Qatar, and Swaziland. This positive correlation between democratic governance and public government consumption is not driven by omitted variables, such as a country's level of economic development.[53]

Military Spending. The relationship between various forms of private government consumption and nondemocratic governance is equally informative. One such expenditure is military spending. While a country's security environment and

---

[49] The government's utility function can also be interpreted as its survival function. The government will choose the optimal "bundle" of public and private consumption that maximizes its probability of staying in power.

[50] This trade-off between the provision of patronage and welfare goods across democracies and autocracies is operationalized in the formal model in the appendix.

[51] Wintrobe (1998), 7–14.

[52] Ghandhi and Przeworksi (2006, 2007) advance theoretical arguments and provide empirical evidence documenting the variation of public and private goods provision within autocracies.

[53] For instance, in a regression that controls for a country's log GDP per capita (averaged over 1990–2015), the coefficient on the average POLITY remains positive and statistically significant. This indicates that governments in more democratic countries tend to allocate more on spending, even when factoring in a country's underlying economic wealth.

3.3 Comparative Public Finances

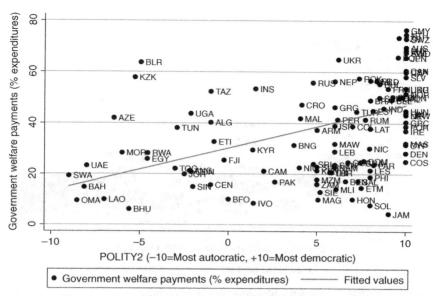

FIGURE 3.3 Government welfare payments and democratic governance.
Notes: The figure plots the average government expenditures on transfers and subsidies (% expenditures) and POLITY score for each country.

wealth (material capabilities) are robust determinants of military spending, regime type is also a substantial factor: democracies tend to allocate less to defense spending.[54] Figures 3.4 and 3.5 provide a snapshot of this relationship: less democratic countries tend to have a higher "defense burden" (total military spending as a share of GDP).[55] Figure 3.4 shows that in a cross section of countries, more democratic countries tend to spend less on defense (as a share of GDP). Figure 3.5 shows that the average defense burden in autocracies has exceeded those in democracies since the mid-1980s. Why does a country's level of democratic governance matter? Because of its greater political benefits in nondemocracies: many dictatorships maintain a large military both as a potential instrument of repression as well as a means to ward off potential challengers to the government.[56]

Corruption. Private government consumption can manifest in other ways, beyond simply entailing direct financial outlays by the government on patronage. For example, the willingness of a government to engage in and/or

---

[54] Dunne and Perlo-Freeman (2003).
[55] The negative relationship between defense burden and democratic governance is not driven by outliers. For example, the negative association holds when omitting potential outliers, such as oil-producing countries in the Middle East (e.g., Saudi Arabia, Kuwait, Qatar, Oman).
[56] Besley and Persson (2010).

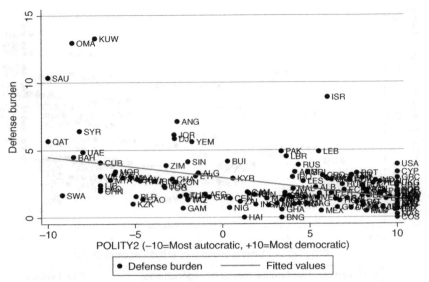

FIGURE 3.4 Defense burden and democratic governance.
Notes: The figure plots the average of defense burden and POLITY score for each country, using all available years of defense spending data.

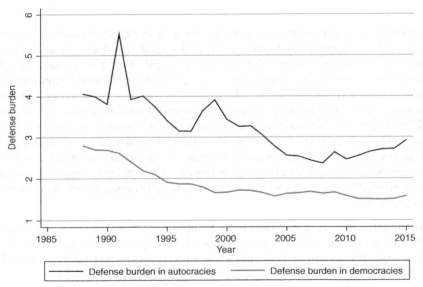

FIGURE 3.5 Defense burden in autocracies and democracies, 1986–2015.
Notes: The figure plots the *annual* average of defense burden across autocracies and democracies. Autocracies are those with POLITY scores less than +6. Democracies are those with POLITY scores equal to +6 or above.

3.3 Comparative Public Finances

permit others to engage in corruption represents a form of private government consumption and corruption tends to be higher in nondemocracies.[57] According to selectorate theory, governments in autocracies are likely to foster corruption for at least three reasons.[58] First, eliminating corruption and encouraging institutions that promote the rule of law are public goods. Dictators have few incentives to find and eliminate corruption because they care less about public goods. Second, leaders can provide benefits by granting the supporters the right to expropriate resources for themselves. Thus, autocratic leaders might encourage corrupt practices as a reward mechanism. Third, the prevalence of kleptocracy in autocracies frequently allows leaders to siphon off resources for pet projects. "Ruling to steal" constitutes a form of corruption. These two latter forms of corruption are consistent with private government consumption that enriches the leader and his key supporters, especially in nondemocracies. Figure 3.6 provides some empirical evidence. The figure plots a country's average level of patronage-based corruption (where a higher value implies greater corruption) against its average POLITY score.[59] It can be clearly seen that corruption is higher in countries with inferior democratic governance.

---

[57] In the extant literature, corruption is usually understood to mean the "misuse of public office for private gain," where the "private gain" may accrue either to the individual public official or to groups or parties to which he or she belongs, such as his or her political party or governing coalition (Bardhan 1997). This definition is quite broad and can capture various forms of corruption, such as payments from private citizens to public officials (bribery) and transfers from the government to key groups (patronage).

[58] Bueno de Mesquita et al. (2003), 102–103.

[59] The conjecture that remittances ease the pressure on governments to deliver public services in order to finance patronage requires a measure of patronage corruption. To measure government corruption in the form of patronage, I use the International Country Risk Guide (ICRG) corruption index. The ICRG's official documentation makes clear that its corruption measure "is more concerned with actual or potential corruption in the form of excessive patronage, nepotism, job reservations, 'favor-for-favors', secret party funding, and suspiciously close ties between politics and business." The assessments of corruption are conducted by a team of ICRG country experts based on available information. The ability of these experts to form their assessments does not depend on changing economic and political conditions for the country under review. For instance, there is no evidence to suggest that periods of domestic unrest (e.g., due to higher commodity prices) garner heightened analysis. Moreover, to the extent that government patronage cannot be reliably quantified for a wide set of developing countries, the ICRG corruption index does capture the trade-off between government patronage and the delivery of government services. For instance, the corruption index is negatively correlated with government expenditures on public health care, education, and social contributions and positively correlated with the size of the public sector (Ahmed 2013). Other scholars (Treisman 2000; Alesina and Weder 2002) have documented that the ICRG corruption measure is highly correlated with corruption measures from alternate sources (e.g., Transparency International) and is also correlated with factors indicative of misgovernance such as public health expenditures.

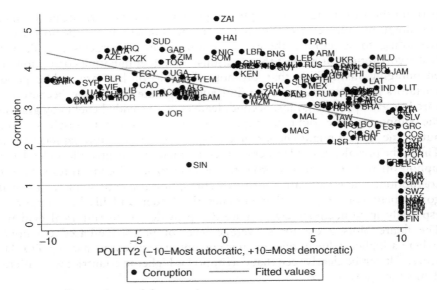

FIGURE 3.6 Corruption and democratic governance.
Notes: The figure plots average corruption and POLITY score for each country. Corruption is the ICRG patronage-based index (1 to 6), where higher values correspond to greater corruption.

Political Survival

For leaders, pursuing repressive tactics and/or cultivating various forms of loyalty (or some combination thereof) is rational if it helps them to remain in power. This is the case in nondemocracies and less so in democracies. To evaluate this, I examine the correlation (ρ) between leader survival (based on the measure of leader termination from Chapter 2) and various measures of repression and loyalty across democracies and nondemocracies.[60] For example, autocrats who govern in countries with increasingly *less* free political rights are less likely to lose power ($\rho = -0.10$), while governments in democracies that engage in such repression are more likely to lose office ($\rho = 0.03$). Strategies of loyalty accumulation also exhibit similar correlations. For example, autocrats who spend more on the military (e.g., with a higher defense burden) are less likely to lose power ($\rho = -0.05$). In contrast, military spending is uncorrelated with political survival in democracies ($\rho = 0.005$). Chapters 4 to 6 provide more robust evidence that repression- and patronage-based measures of loyalty (e.g., corruption, military spending, public sector employment) strengthen autocratic rule.

---

[60] A negative correlation implies the leader (government) is less likely to exit from political office.

3.4 INTERNATIONAL CAPITAL AS NONTAX INCOME

Channels

These political foundations of government spending provide a framework to theorize how international capital can affect political survival. Foreign aid, remittances, and FDI represent financial transfers that governments, households, and firms can use in a variety of ways. For instance, a government that receives $100 million in foreign aid can invest it in public infrastructure projects, pay down government debt, and/or "steal" some of it. Households that receive remittance income from abroad, for example, may save it, choose to invest these funds in their children's health and education, or simply spend the funds on goods and services. FDI embodies an infusion of capital – often accompanied with a transfer of new technology and human capital – that can help domestic firms become more productive. According to standard economic theory, these productivity gains should lead to higher wages, but they may also generate disproportionate benefits (rents) to the owners of the firms (e.g., elites, the government via state-owned enterprises), as well as individuals (e.g., family members of the leader) or groups (e.g., the military) that are affiliated with these owners. Depending on how governments spend foreign aid, households consume their remittances, and firms allocate their rents and productivity gains from FDI, governments can strategically position themselves to augment their public finances to ensure their political longevity. Doing so may be especially attractive for an autocrat.

An autocrat could incorporate these three distinct types of international capital into its revenue base through a sequential process or "game." In the first stage, firms and households optimize their decisions accordingly (e.g., the bribes firms might pay, the spending decisions of households with remittance income). In the second stage, the government "observes" the behavior of firms and households and decides its "best response" to maximize its revenue in order to optimally allocate its spending on repression and loyalty. I describe these channels in the text that follows.

Income Effect. Foreign aid is a direct transfer from a donor government (or multilateral aid agency, such as the World Bank) to a recipient government. Thus, foreign aid will directly enter an autocrat's revenue base and the autocrat is free to optimally allocate that aid between private and public government consumption. Foreign aid therefore has a direct "income effect" on government spending. In contrast, remittance income and FDI do not directly accrue to an autocrat's revenue base. This does not imply, however, that these flows cannot be integrated into an autocrat's revenue base through alternate channels.

Substitution Effect. Remittances represent foreign income for households. Thus, by definition, remittances increase household income, which may be partially extracted by a government via taxation. Yet extant evidence shows that developing countries often lack the capacity to monitor, track, and effectively tax remittance income.[61] As such, remittances are often viewed as a form of nontax household income.[62] This does not mean that remittance cannot be "taxed" via an alternate mechanism. For example, at the margin, governments may be able to adjust their expenditures between private and public consumption in response to an increase in remittance income.

This adjustment in government spending would occur in the second stage of the game. In particular, additional remittance income may permit a government to reduce its expenditure on certain welfare goods (e.g., health care, public education) and divert those "freed" funds for patronage (e.g., corruption, government salaries) because a remittance-receiving household may have adequate funds to make up the difference.[63] Thus, remittance income can substitute for a reduction in public government consumption. (This substitution effect associated with remittances is formalized in the appendix to this chapter.)

Taxation and Rent Creation. Foreign direct investment may also affect a government's public finances. One channel is FDI increasing a firm's output and productivity, thereby raising labor wages.[64] Higher wages are taxable. Higher wages may also strengthen workers' affinity for the government and help legitimize a government's authoritarian policies. FDI may also enter a government's revenue base through an alternate, nontax channel. Foreign investment can create opportunities for extractive rents, especially in industries with high entry barriers and fixed costs, such as petroleum, chemicals, shipping, and steel.[65] For instance, foreign firms may pay bribes to bureaucrats or other government affiliates (such as military personnel) to win government contracts. Alternatively, the government may permit foreign companies to enter joint ventures only in industries and/or with companies that are affiliated with supporters of the regime. As such, the profits generated by the joint venture will disproportionately go to regime supporters. As examples, alliances between the military, bourgeoisie, and foreign capital underpinned the durability of many bureaucratic-authoritarian regimes in the 1970s to 1990s, such as those of the "Triple Alliance" in Brazil and the "New Order" in Indonesia.[66]

---

[61] de luna Martinez (2005). [62] Chami et al. (2008). [63] Abdih et al. (2012).
[64] In standard models of firm production, a higher marginal productivity of labor (due to an infusion of productive capital) leads to higher wages. This is demonstrated in the appendix.
[65] These opportunities for rents tend to be especially high in industries with high entry barriers. These industries tend to require fixed assets and predominantly immobile investments, leading to a fewer number of firms and greater opportunities for monopoly or oligopoly rents (Pinto and Zhu 2016).
[66] On bureaucratic-authoritarianism, see Wintrobe (1998), 13. On Brazil and Indonesia, see Evans (1979) and Robison (1988), respectively.

3.4 International Capital as Nontax Income

TABLE 3.2 *Channels*

Capital inflow	Channel	Government behavior
Foreign aid	Income effect	Directly finances private government consumption
Remittances	Substitution effect	Reduce welfare goods provision and diverts expenditures to private government consumption
FDI	Rent creation	Additional tax revenue (via higher wages) and opportunities for rents directly finances private government consumption

The Cumulative and Magnified Effect in Nondemocracies

Using these three distinct channels (summarized in Table 3.2), governments have the means to incorporate foreign capital into their revenue base and better ensure their prospects for political survival. Of course, a country's underlying institutional environment can affect how these rents are spent and their efficacy (utility) for governments. In particular, when there is smaller winning coalition to whom rents are transferred, the incentives and payoffs for governments to incorporate international capital into their revenue base become *greater* in autocracies.

This can be illustrated with a simple example. Suppose a country has a population of 40 million and inflows of international capital equal to $500 million (e.g., $250 million in FDI, $150 million in remittances, and $100 million in foreign aid). Suppose a government could at most incorporate 40 percent of the inflows into its revenue base (representing $200 million in potential rents) via the mechanisms discussed in the preceding subsection.

Now consider the spending decisions and relative payoffs for governments in a democracy and an autocracy. Suppose an incumbent democratic government must retain the loyalty of at *least half* of the country's population with targeted benefits (e.g., recall Figure 3.3, which shows that democratic governments spend a greater share of their budgets on welfare programs). This means the incumbent transfers $10 in rents to half of the country's citizens. Such a transfer is relatively small and may not necessarily guarantee the loyalty of each of the recipients to the government. In contrast, the same capital inflow can generate a stronger political dividend for an autocrat. Suppose an incumbent autocrat needs to retain the loyalty of the military, which comprises 5 percent of the population. This means an autocrat can transfer $100 to each military officer – an amount that is more likely to retain the loyalty of an officer (and which is substantially higher than the transfer to a citizen in a democracy).

This simple example highlights two important points. First, with a smaller winning coalition, autocrats can disburse higher per capita transfers to its pivotal regime's supporters. Second, because these *foreign* rents generate greater political utility for incumbents in nondemocracies, autocrats have a greater incentive to engage in the mechanisms (outlined earlier) to incorporate foreign aid, remittances, and FDI into their revenue base. Thus, a country's underlying political institutions and inflows of foreign capital can incentivize and benefit autocrats more so than democrats.

This generates two implications that Chapters 4 to 6 evaluate empirically:

1. *Channels.* A government can incorporate different types of international capital into its revenue base via three distinct channels: foreign aid through an income effect, remittances through a substitution effect, and FDI through a rentier effect.
2. *Political survival.* In doing so, these capital flows constitute a source of nontax government income (rent) that a government can spend on strategies of political survival: repression and accumulating loyalty. A country's institutional environment will affect the relative weights with regard to the types of government spending. In polities with a smaller winning coalition (i.e., nondemocracies), leaders will allocate a greater share of the government's nontax revenues (stemming from international capital) on funding repression and patronage, given that the payoff from doing so is greater in nondemocracies. This means that international capital is especially beneficial to the political survival of dictators, and less so for democrats. This generates a clear empirical implication: if dictators are optimizing, then in equilibrium, additional international capital should *not* accelerate the government's removal from power.[67]

The discussion in this section, which is summarized in Table 3.2, provides a nontechnical description of how foreign aid, remittances, and FDI can augment a government's revenue base to fund strategies of political survival. For more technically inclined readers, the appendix to this chapter advances a parsimonious model formalizing these channels in a game-theoretic framework. Motivated by the nontechnical description of a two-stage game in the preceding subsections, the model is a two-stage sequential game, where firms and households are first-movers. In a few simple steps, the model shows how leaders in *increasingly* autocratic polities are incentivized to use foreign aid, remittances, and foreign investment to fund repression and accumulate loyalty (through patronage).

---

[67] As I will show in the empirical chapters, this means that *additional* capital inflows will either lower the probability of leader "exit" *or* have no effect *in dictatorships*. The latter is a "weaker" finding of international capital on authoritarian survival. It is worth stating that *if* additional capital *increases* the probability of leader (government) termination *in nondemocracies*, this would weaken the claims in this book.

3.5 THEORETICAL AND EMPIRICAL CONCERNS

Potential Countervailing Effects

While this chapter has developed a revenue-based theory of international capital and authoritarian rule, it is plausible that international capital may undermine autocracy through countervailing channels. For example, all three types of foreign capital can raise individual incomes and thus cultivate conditions for pro-democracy sentiment consistent with "modernization theory."[68] And separately, each type of capital has the potential to affect autocratic politics through channels unrelated to a government's public finance.

For instance, many donor governments frequently allocate foreign aid to "reward" countries that pursue "good governance" and undertake democratic reforms (e.g., reduce corruption). While it is difficult to "quantify" these policy inducements, they may nonetheless undermine autocracy and offset the income effect associated with foreign aid advanced in this chapter. Relatedly, in an effort to cultivate an investor-friendly climate, governments often undertake regulatory reforms that move to liberalize domestic politics (e.g., increase transparency). As a consequence, inward FDI may generate a countervailing effect on autocracy that is independent of a rentier effect. Identifying an exogenous source of variation for the capital flow (e.g., that is uncorrelated with policies promoting "good governance") is a means to circumvent the allocative decisions of donor governments and "sending" firms.

Finally, it is plausible that foreign capital can shape a government's public finance through channels that do not feature in this chapter's theory. For example, in addition to shifting the composition of government spending (i.e., via a substitution effect), higher levels of aggregate remittances could expand *total* government spending by reducing the cost of sovereign borrowing. This potentially complementary mechanism relies on the notion that remittances help ease sovereign borrowing costs by enhancing a country's creditworthiness.[69] By facilitating a country's access to international credit (and to greater accumulation of foreign assets), remittances may allow a government to expand its total spending, including greater spending on patronage.[70]

While these countervailing effects do not necessarily weaken the veracity of the theory developed in this chapter, they do pose challenges to empirical evaluation. Cognizant of this concern, the empirical analysis in Chapters 4 to 6 discusses these potential countervailing effects and presents evidence to discount their empirical effects on nondemocratic politics.[71] In addition, for

---

[68] Lipset (1959). [69] Ratha (2007).
[70] Singer (2012) develops this argument. Cognizant of this alternate channel, in Chapter 5, I provide stronger evidence in support of the substitution effect mechanism over the sovereign borrowing mechanism.
[71] For example, because all types of foreign capital can affect economic development, the empirical analysis controls for each recipient country's economic growth and per capita GDP (country

each type of capital flow, the empirical analysis will show its causal impact on institutions, governance, *and* political survival.

This, of course, is not the only concern in empirically evaluating the causal veracity of the theory developed in this chapter.

Causal Identification and Evaluating Channels

Empirically evaluating whether each capital inflow *causes* authoritarian governance via the channels developed in this chapter is challenging. The principal worry is endogeneity: capital inflows are correlated with domestic politics in recipient countries and vice versa. For instance, donors may disburse foreign aid to accelerate political reform in democracies, in which case countries with "good institutions" (less autocratic) may receive higher amounts of aid.[72] Some donors, however, may also use aid to buy policy concessions, especially from autocracies where the costs of pursuing such a strategy is lower due to a smaller winning coalition to buy off.[73] Thus, donors may disburse higher amounts of aid to less democratic countries. Domestic politics in sending states can also be an important determinant of international migration and remittance inflows. For example, poverty is a main driver of emigration, and more impoverished countries are frequently plagued by weak institutions and nondemocratic governance.[74] Finally, domestic political conditions, such as the constraints faced by national leaders and the quality of property rights protection, are an important factor in explaining cross-national FDI.[75] These factors are correlated with the quality of democratic governance in the investment receiving country.

Addressing these concerns with endogeneity and establishing a causal relationship is challenging. In the past twenty years, social scientists have developed a number of approaches to "identify" causal relationships in observational, real-world data. At the heart of these approaches is the quest to identify a *plausibly exogenous* source of variation for the endogenous variable (e.g., the relevant international capital inflow) that is uncorrelated with the outcome variables of interest (e.g., political institutions, leadership failure, government expenditures, corruption).

In Chapters 4 to 6, I tackle these issues head-on. As each type of capital flow can affect authoritarian politics via a distinct channel, I devote a separate chapter for each capital flow. In each chapter, I evaluate the causal impact of that capital flow on authoritarian governance (e.g., its institutions, composition

wealth). The analysis also controls for time-invariant characteristics (e.g., colonial relationship, distance) that may affect aid, remittance, and FDI flows to recipient countries. For instance, donors often allocate greater amounts of foreign aid to former colonies. Regressions that include country fixed effects will account for this pattern of aid allocation.

[72] Finkel et al. (2007).   [73] Bueno de Mesquita and Smith (2009).   [74] Chami et al. (2008).
[75] Li and Resnick (2003).

of government spending) *and* political survival. Moreover, by focusing on a specific type of capital, each chapter allows me to better evaluate whether that capital flow affects authoritarian politics through its distinct channel.

I evaluate these channels using a common approach in political economy: by testing whether the independent variable of interest exhibits a statistically significant effect on a dependent variable that is consistent with the proposed theoretical channel. For example, Chapter 5 evaluates the impact of remittances on both institutions (executive constraints) and political survival and also tests whether this occurs through a substitution effect (as argued in this chapter). To hone in on this proposed channel, Chapter 5 presents evidence that remittances lower a government's provision of welfare payments (i.e., a measure of public goods spending) while increasing expenditures on public sector employment and patronage-based corruption (which are both measures of private goods spending in autocracies).

An alternative approach to evaluate this channel could be mediation analysis, which would gauge whether the proposed channel is a "mediating variable" that effectively nullifies the statistical relationship between the key independent and dependent variables. For example, in Chapter 5 this means that "controlling" for welfare payments (or expenditures on public sector employment) in a regression model would significantly weaken any statistical relationship between remittances and both executive constraints and political survival. While valid, this approach is a very stringent test of causal channels, as it implies that the substitution effect is the only channel through which remittances affect authoritarian politics. I do not make such bold claims in this chapter. Rather, I present a framework identifying potential channels through which various types of international capital can affect political survival. It is quite plausible that international capital can affect political survival through additional channels, such as higher economic growth. I do not rule these out as possible theoretical channels. Rather, as discussed in the previous subsection in the analysis to follow, I carefully evaluate these potential alternate channels and present evidence to discount their empirical validity.

3.6 SUMMARY

Starting from microfoundations, this chapter advances a revenue-based theory of international capital and authoritarian rule. The chapter begins by articulating the dual problems of authoritarian rule (in Section 3.1) and hones in on the critical role of government revenue – particularly, its nontax sources – in helping autocrats mitigate these challenges. In Section 3.2, I argue that with sufficient revenue a government is able to finance two broad classes of strategies that better ensure its political survival: repression and loyalty. The latter can be strengthened through supplying patronage and/or cultivating legitimacy. In Section 3.3, I describe how political institutions affect the composition of government finance. Here, I differentiate between public and private government consumption (on public welfare vs. military expenditures as examples). In nondemocracies, governments are incentivized to allocate a greater share of their expenditures on financing political repression and targeted transfers to key groups, such as the military and regime supporters with employment in the public sector.

These building blocks underlie a strategic framework through which international capital can finance authoritarian rule. This strategic process – or "game" – is described in Section 3.4 and is formalized in the chapter's appendix. The game is sequential and involves two stages. In the first stage, firms and households are the recipients of FDI and remittances, respectively. Firms and remittances optimize their behavior accordingly. FDI raises the productive capability of firms, generating higher profits that can be taxed by the government and/or used to pay bribes to the government.[76] Migrant households can spend their remittance income on consumption goods (e.g., food, clothes) and substitutable welfare goods (e.g., health care, education).

In the second stage, the government "observes" the behavior of firms and households. The additional tax income and rents (bribes) from firms – associated with FDI – directly accrue to the government. This constitutes a *rentier effect*. In contrast, the government is able to incorporate remittance income through a *substitution effect*. The decision of migrant households to spend more on welfare goods allows the government to spend less on this type of good and to increase its expenditures on targeted transfers to its supporters (patronage) and/or to finance political repression (e.g., expenditures on the government's security apparatus). Finally, foreign aid is a transfer directly to a government that it can spend however it likes. This comprises an *income effect*. Thus, through a rentier effect (associated with FDI), a substitution effect (associated with remittances), and an income effect (associated with foreign aid), a government is able to fund the strategies that better ensure their political longevity. As discussed in this chapter, in *nondemocracies*, governments are incentivized to spend more on repression and patronage. Thus, FDI, remittances, and foreign aid can finance authoritarian rule. The chapter concludes in Section 3.5 by discussing potential theoretical and empirical challenges that subsequent chapters will address.

---

[76] Greater firm profits may also translate to higher wage compensation to employees and/or higher dividend payments to shareholders. This is made explicit in the formal model.

Appendix: Model

APPENDIX: MODEL

Overview

In this appendix, I present a parsimonious, single-shot game-theoretic model that illustrates how an increasingly autocratic government (leader) can act strategically to incorporate foreign aid, remittances, and FDI into its revenue base. As a result, the autocrat can fund strategies that enhance his or her prospects of political survival: repression- and patronage-based loyalty. The model builds on the insights of several existing theoretical formulations linking public finance and institutions to political survival, including Bueno de Mesquita et al. (2003), Acemoglu and Robinson (2006), Abdih et al. (2012), and Ahmed (2012). The innovation of my model is the integration of three distinct forms of international capital to three different actors in an economy: foreign aid to governments, remittances to households, and foreign investment to firms. In a few simple steps and employing some basic game theory, the model derives the channels associated with each capital flow, as depicted in Table 3.2.

The construction of the model is based on arguments in the chapter. First, government revenues are the foundation of any government action (Section 3.1) to fund repression and/or accumulate loyalty (Section 3.2). Second, the quality of democratic institutions affects how governments spend their revenues on private government consumption (repression, patronage) and public government consumption (welfare goods). Specifically, governments in nondemocratic polities will place greater "weight" on private government consumption relative to public government consumption (Section 3.3). Third, because foreign aid, remittances, and FDI are received (primarily) by three distinct actors in a country, the model has three actors: firms, households, and a government (Section 3.4). FDI increases a firm's capital stock, remittance enrich household income, while foreign aid flows directly to a recipient government's coffers.

With these building blocks and using some basic game theory, in a few steps, I show how governments can strategically incorporate these capital inflows into their revenue base.[77] They do so via three distinct channels. First, there is an income effect whereby some fraction of foreign aid is directly siphoned to pay for private government consumption. Second, there is a substitution effect whereby remittance income allows a government to reduce its provision of welfare goods (that would benefit households) and divert those expenditures toward private government consumption. Third, foreign investment raises domestic wages that a government can directly tax. Foreign investment may

---

[77] The functional forms of the utility functions for the agents (firms, households, and the government) and nature of strategic interaction among these agents follows the setup in Abhih et al. (2012) and Ahmed (2012).

also create opportunities for outright extraction (rents) by a government. Additional tax income and rents can finance private government consumption. In equilibrium, the model shows that these channels are magnified in countries with more autocratic institutions, whereby the combination of foreign aid, remittances, and foreign investment allows autocrats to finance policies of repression and patronage to ensure their political survival.

Players and Preferences

A. Firms

There is a representative firm with a Cobb–Douglas production with three inputs: foreign capital (F), domestic capital (K), and labor (L). The firm's output is given by

$$Y = (F+K)^\rho L^{1-\rho} - D \tag{3.A1}$$

where D are the potential rents a government can "steal" – for example, through bribes paid by firms and rents created by joint ventures between multinational firms and state-owned enterprises. D could vary by a country's underlying quality of democratic governance: the incentive to steal is higher for governments in nondemocratic polities. The parameter ρ is the weight firms place on capital relative to labor ($0 < \rho < 1$).

Firms maximize profit subject to their cost of production (C):

$$C = wL + m(F+K) \tag{3.A2}$$

where w is the wage rate and m is the rental rate for both domestic and foreign capital.[78] Firms set wages subject to their profit maximization problem.[79] The first-order conditions from this maximization problem imply that the marginal product of labor equals the wage rate:

$$w^* = \frac{(1-\rho)}{\rho} \frac{(F+K)}{L} \tag{3.A3}$$

Equation 3.A3 shows that the optimal wage paid by firms increases with additional foreign capital (F) and domestic capital (K) and is inversely related to labor supply (L) and the capital to labor share (ρ). In particular, as firms become more productive from an additional unit of foreign capital, these productivity gains translate to higher wages for workers. Specifically, $\frac{\partial w^*}{\partial F} > 0$.

---

[78] For simplicity, I assume that the rental rate is the same for domestic and foreign capital, although the substantive implications are unchanged if one assumes a different rental rate for domestic and foreign capital.
[79] Firms maximize profits by maximizing their output (given by equation 3.A1) subject to their total cost (given by equation 3.A2).

Appendix: Model

B. Households

For households, the wages paid by firms help finance the consumption of two types of goods. The first is a purely private good purchased exclusively by the household. The second is a "substitutable" welfare good that can be provided by either the government or the household (e.g., education, health services).[80] The marginal utility of this welfare good does not depend on who pays for it, although households would prefer the government to supply this good.

Households have Cobb–Douglas preferences over these two types of goods, given by the log utility function:

$$U(c, p, g) = \gamma \log(c) + (1 - \gamma) \log(p + g) \qquad (3.A4)$$

where c is the representative household's consumption of the private good, p is the household's consumption of the welfare good, and g is the government's provision of that welfare good. The parameter γ is the weight households place on private goods relative to welfare goods ($0 < \gamma < 1$). Households finance their expenditures subject to their budget constraint:

$$(1 - \tau)w + R = c + p \qquad (3.A5)$$

where w is the household's income (from wages only), τ is the tax rate, and R represents remittances. R is untaxed by the government.

C. Government

Governments also care about private goods (patronage) and public goods (welfare), but do so in relation to their main objective of staying in power. They do this by redistributing economic and/or political rents to key individuals (e.g., party supporters, business elites, military officials) and groups (e.g., coethnics of the leader/government, organized labor, the majority of the voting population) in return for their political support. The distribution of rents can be in the form of welfare (e.g., subsidies and transfers to the general public) and patronage goods (e.g., corruption, public sector compensation). The relative importance of providing private government goods (s) to welfare goods (g) is captured by α in the government's utility function:[81]

$$\pi(s, U) = \alpha \log(s) + (1 - \alpha) U(c, p, g) \qquad (3.A6)$$

where $0 < \alpha < 1$ and s is whatever the government keeps for its own private consumption. This private government consumption can finance repression and maintain loyalty, via various forms of patronage. *In more authoritarian leaning polities, α is higher.* For example, α might be very close to 1 in North Korea

---

[80] This theoretical formulation builds on the work of Abdih et al. (2012) and Ahmed (2012).
[81] The government's utility function can also be interpreted as its survival function. The government will choose the optimal "bundle" of patronage and welfare goods to maximize its probability of staying in power.

because the reigning dictatorship spends very little on the welfare of its population.[82] In democracies, α will be closer to zero.[83]

Given these parameters, the government chooses s to maximize its utility function subject to its budget constraint:

$$\tau w + A + D = g + s \qquad (3.A7a)$$

and

$$g^* \geq g_{\min} \qquad (3.A7b)$$

where τ is the tax rate, w is income, A is foreign aid (or any type of nontax government income, such as income from sovereign borrowing or natural resources), and D are rents associated with foreign capital (given by equation 3.A1). Equation (3.A7b) ensures a government satisfies a revolutionary constraint, whereby a government needs to spend a minimum amount on welfare goods to prevent rebellion.[84]

Sequence of Moves

There are three actors in this simple economy: firms, households, and a government. And each actor is the recipient of a distinct type of foreign capital: FDI to firms, remittances to households, and foreign aid to the government. These capital flows in turn affect their optimal decisions (as shown earlier). To model the interaction of firms, households, and the government, I consider a sequential process with the following series of moves:

First, firms and households optimize their decisions. Firms choose their optimal level of wage to pay workers and "bribes" (i.e., rents to the government). Households choose their optimal levels of spending on private and welfare goods.

Second, the government optimizes its spending on private and public government consumption.

As the game is sequential, deriving the government's "equilibrium" behavior requires taking into account the optimal decisions of firms and households (who made their decisions *before* the government). Solving for the equilibrium, therefore, requires backward induction.

---

[82] For example, the North Korean government often leverages its nuclear technology/weapons program to "extract" concessions from other countries in the form of food aid because the government undersupplies food to a large share of its population.

[83] Empirically, α could be conceptualized (measured) as an inversion of the POLITY score. Specifically, POLITY scores closer to −10 correspond to values of α closer to 1, while POLITY scores closer to +10 correspond to values of α closer to).

[84] Acemoglu and Robinson (2006).

Appendix: Model

Solving the Game through Backward Induction

To keep the analysis simple, I model the interaction between household and government as a one-shot Stackelberg game where the government moves first. This means the government determines its optimal spending decisions in response to the household's optimal consumption choices, i.e., a government acts strategically in response to household behavior. Solving for the equilibrium therefore requires backward induction.

This first step requires determining the household's optimal consumption of welfare goods – and by extension, consumption of private goods – by maximizing its utility function (with respect to p) subject to its budget constraint. Doing so yields the household's optimal provision of welfare goods:

$$p^* = (1-\gamma)[(1-\tau)y + R] - \gamma g. \quad (3.A8)$$

Equation (3.A8) shows that the higher amounts of total household income (which consists of after-tax income and remittances) increases a household's optimal expenditures on welfare goods. Equation (3.A8) also shows in response to a provision of the welfare good by the government (i.e., a higher value of g), a government will decrease its own expenditures on the welfare good.[85] Since the household allocates its budget between p and c, if household expenditures on the welfare good rise then expenditures on private goods must decline (and vice versa).

In the first stage of the game, the government incorporates the household's optimal provision of welfare goods in its survival function. After making this substitution, the government's optimal provision of welfare goods is determined by maximizing its survival function (with respect to g) subject to its budget constraint. The government's optimal provision of welfare goods is given by

$$g^* = (\tau - \alpha)w + (1-\alpha)(A+D) - \alpha R. \quad (3.A9)$$

Equation (3.A9) demonstrates that higher wages (w), foreign aid (A), and rents from foreign investment (D) increase a government's expenditures on welfare goods. In contrast, higher levels of remittances (R) induce a government to spend less on welfare.

Whatever the government does not spend on welfare goods can be spent on private government consumption. Specifically,

$$s^* = \alpha(A + D + R + w). \quad (3.A10)$$

Equation (3.A10) clearly shows that *higher amounts of foreign aid, rents from foreign investment, and remittance income increase a government's*

---

[85] The model assumes that the government can approximate the amount of aggregate remittances. For instance, governments (even in developing countries) could approximate these inflows based on publicly available data from the World Bank and the International Monetary Fund.

optimal expenditures on its private consumption. Foreign investment also has an indirect effect on s via wage income (w) because additional foreign investment can translate to higher firm productivity and higher wages (given by equation 3.A3). Because the sum of foreign aid, remittances, and foreign investment inflows comprises international capital (as conceptualized in this book), it follows that s^* *is increasing in international capital.*

Finally, equation (3.A10) shows that this overall effect is *magnified* in countries with *higher levels of autocratic governance,* α. Stated plainly, higher amounts of international capital received in *more* autocratic countries finance greater private government consumption, thus strengthening the survival prospects of governments in autocracies.

Interpretation

A. The Channels

Given the incentives of authoritarian regimes to engage in patronage and repression, nondemocratic governments will place greater weight on expenditures on private government consumption (compared to democracies) in their survival function. This means α is higher in autocracies than in democracies, where values of α closer to 1 correspond to more authoritarian politics. This interpretation generates three channels associated with each type of international capital flow.

First, a government's optimal provision of welfare goods (given by equation 3.A9) depicts a *substitution effect* associated with remittances and welfare good provision, and this effect is *magnified* in countries with *more* autocratic politics (i.e., $-\alpha R$). Plainly stated, governments in countries with more authoritarian politics will tend to reduce their provision of welfare goods the most.

Second, equation (3.A9) also shows that governments spend some fraction of foreign aid, $(1 - \alpha)A$, on welfare goods, while the rest (αA, given by equation 3.A10) is spent on private government consumption for repression and/or the provision of patronage goods. In more autocratic polities, a lower fraction of aid is spent on welfare goods because α is larger (i.e., closer to 1). Consequently, a higher share of aid is spent on repression and patronage in countries with more authoritarian politics. This constitutes an *income effect.* Similarly, potential rents associated with foreign investment can finance patronage and repression (i.e., αD). This represents a rentier effect.

Third, equation (3.A10) shows that greater amounts of foreign investment can increase the government's private consumption directly via a *rentier effect* (i.e., αD) and indirectly through higher wage income (i.e., αw). This latter channel is consistent with the notion of "performance legitimacy," whereby higher *earned* household income can bolster a government's tenure in power.

Thus, foreign aid, remittances, and foreign investment can affect authoritarian politics via three distinct channels: a *substitution effect* (associated with remittance income), an *income effect* (associated with foreign aid), and a *rentier effect* (associated with foreign investment). There is also an indirect "performance legitimacy" channel associated with higher wages via foreign investment.

The combination of these channels frees resources for a government to increase expenditures on private government consumption. Equation (A.310) shows this explicitly. It demonstrates that a government's private consumption is increasing in autocracy (a), foreign aid, remittances, and foreign investment. Moreover, the overall effect is *magnified* in countries with higher levels of authoritarian governance, i.e., $a(A + D + R + w)$. As s^* represents expenditures an autocrat can use to finance his or her various possible strategies of political survival (e.g., repression, patronage, corruption, military spending), equation (3.A10) clearly demonstrates that an autocrat's prospect of survival is increasing in the interactive effect of autocracy and international capital. This conjecture is consistent with the relationships in Figures 2.4 to 2.6 from Chapter 2. Chapters 4 to 6 will examine each capital flow separately and its associated causal channel (e.g., substitution effect associated with remittances, rentier effect associated with foreign investment).

B. Welfare Implications

It is also apparent from equation (3.A10) that the government's incentive to divert resources to its own private consumption, measured by a, leaves the household worse off in equilibrium. This can be assessed by examining how a household's utility changes as a country becomes more authoritarian. Specifically, replacing equations (3.A8) and (3.A9) into (3.A4) yields

$$\frac{\partial U(c^*, p^*, g^*)}{\partial a} = -\frac{1}{(1-a)(A+D+R+w)} < 0. \quad (3.A11)$$

Equation (3.A11) shows that international capital received in a country with *more* authoritarian politics has an *increasingly* larger negative effect on household welfare via the quality of a country's underlying politics (e.g., formalized institutions).[86] This does not necessarily imply that international capital has a net negative welfare effect on household welfare. For example, from equation (3.A8), it still remains true that international capital has a positive effect on household utility through higher consumption on private goods financed by higher wages (from foreign investment) and remittance income. Moreover, from equation (3.A9), it is clear that some fraction of aid is spent on the government's provision of welfare goods, which raises household utility (via equation 3.A4).

---

[86] For example, suppose international capital is held "fixed" at 2 and consider two values of $a = 0.2$, 0.5. Under $a = 0.2$ and with international capital equal to 2, equation (3.A11) yields a value of -0.625. If $a = 0.5$, equation (3.A11) yields a value of -1.

4

Aiding Repression

A central argument in Chapter 3 is the proposition that autocrats can use nontax income to finance a combination of repression and patronage as a means of extending their tenure in office. While various scholars have investigated the relationship between foreign aid and patronage, there is surprisingly less scholarship evaluating the impact of aid on repression.[1] With this in mind, this chapter presents robust evidence that foreign aid from the world's largest bilateral – the United States – can harm political rights, expand the powers of dictators, and entrench nondemocratic institutions in recipient states. As this chapter describes in greater detail, one such example is General Siad Barre's use of US aid to finance his repressive dictatorial rule.

Yet, evaluating whether foreign aid has a causal effect on politics in a broader cross-national setting is challenging, owing in large part to endogeneity. For instance, a recipient's political characteristics often influence the US government's aid allocation decisions. On the one hand, US aid may reward countries committed to political liberalization, such as US aid to shore up nascent Eastern Europe democracies after the end of the Cold War. On the other hand, aid may help stabilize autocratic allies (e.g., Egypt) and thus undermine political liberalization.

To overcome this challenge, I leverage an instrumental variables (IV) strategy. The research design builds on the institutional foundations of US aid decisions, in which the funding and allocation of bilateral economic aid involve both the executive branch and Congress. Congress, in particular, is legally responsible for determining the aid budget. The IV strategy exploits plausibly exogenous variation in the legislative fragmentation of the US House of Representatives ($FRAG_t$) interacted with the probability a country receives US aid (P_i) as a powerful instrumental variable for US bilateral aid to around 150 countries.

Armed with this instrumental variable and controlling for prevailing explanations for political liberalization (such as economic development and

---

[1] Wright (2009), Ahmed (2012), and Morrison (2015) evaluate the relationship between foreign aid and patronage.

growth; time-invariant country characteristics, e.g., colonial legacy; and temporal effects, e.g., Cold War, War on Terror), the IV results demonstrate that US economic aid harms political rights, expands the powers of government leaders, and makes authoritarian institutions more durable.[2] And US aid tends to engender these effects by reducing government tax effort, thus weakening the trade-off between government accountability and taxation.

I proceed as follows. Section 4.1 discusses the politics of foreign aid, honing in on Somalia. This case vignette illustrates how US economic aid financed Siad Barre's repressive reign in Somalia. The section ends with a discussion of how endogeneity muddles causal evaluations of US aid on politics. To tackle this challenge, Section 4.2 discusses the empirical strategy to identify the causal effect of US economic aid on repression. Using this identification strategy, Sections 4.3 to 4.5 present the main results. In Section 4.3, I show that US aid causes governments to lower their quality of political rights. Section 4.4 provides compelling evidence of a channel: foreign aid allows a government to exert less "tax effort," which reduces its incentive to be politically accountable to its population (an argument advanced in Chapter 3). Section 4.5 explores the impact of US aid on leader survival: it shows that aid-induced repression does not undermine leadership tenure in dictatorships. Together, the evidence in Sections 4.3 to 4.5 demonstrate that US foreign aid strengthens authoritarian institutions and does not threaten the political survival of leaders.[3] Section 4.6 concludes.

4.1 FOREIGN AID AND POLITICS

The "Varieties" of Foreign Aid

Foreign aid is quintessentially a transfer of resources (capital, goods, or services) from one government to another. This broad definition covers numerous "varieties" of foreign aid. Foreign assistance can vary in its form and intended purpose, such as *financial* grants and loans for economic development (e.g., investments in roads, schools), military and defense spending (e.g., procuring weapons systems), civil society (e.g., "democracy-assistance" projects), public health (e.g., programs to combat HIV/AIDS), among many others. Foreign aid can also include the transfer of goods, such as food (e.g., shipments of wheat and grains) and weapons (e.g., arms imports), as well as the transfer of services, such as technical assistance to help manage the economic programs and the training of soldiers.[4]

---

[2] The IV specifications (e.g., using aid disbursements on an annual basis) follow the standard approach employed in existing studies, such as Scott and Steele (2011), Wright (2009), and Nunn and Qian (2014).

[3] As noted in Chapter 1, if autocrats are optimizing their decision-making, *in equilibrium*, additional international capital (e.g., foreign aid) should not imperil their political fortunes.

[4] See Alesina and Dollar (2000) for an overview of the determinants of foreign aid.

Foreign aid can also vary in its temporal dimension. For example, in studying the effect of aid on economic growth, Clemens et al. (2012) disaggregate aid by type. They distinguish between short-term aid (e.g., budget support, infrastructure investment, agricultural and industrial support), which can plausibly increase growth in the near term, and other types of aid such as disaster relief (which tends to be negatively correlated with growth) and education spending (which tends to be correlated with long-term growth and less so with short-term growth). Another dimension relates to "donor conditionalities." As implied by its name, conditional aid specifies how recipient governments should use the aid. Conditional aid specifies what the aid can be spent on, such as the specific programs it should fund and/or the particular goods it should purchase from specific donors (e.g., weapons, food, pharmaceuticals). Conditional aid often also delineates the "concessions" a recipient government must make to receive the aid. These can include domestic policy reforms (e.g., trade liberalization, curtailing inflation, cutting government expenditures in specific sectors), as well as policy concessions to donors, such as the donor's access to military bases in the recipient state and currying favor from the recipient government in international organizations (e.g., aligning its voting with that of the donor government in the United Nations).[5]

Policy concessions coincide with two related aspects of foreign aid: donor type and preferences. While most aid involves transfers between sovereign governments, foreign aid can also be disbursed from multilateral organizations such as the World Bank, United Nations, and regional development banks. Unsurprisingly these donors have different preferences that can vary by intended purpose, across recipients (e.g., former colonies vs. noncolonies), and over time (e.g., Cold War vs. post–Cold War period). With respect to purpose, for example, the World Bank's charter states its primary objective in disbursing aid is to facilitate economic development in recipient (member) states.[6] In contrast, bilateral donors may have a stronger desire to achieve their geopolitical objectives with their aid disbursements.[7]

---

[5] Bueno de Mesquita and Smith (2009) provide a political economy framework to understand the relationship between donors and policy concession from recipients.

[6] This does not mean that "politics" cannot affect the allocation of multilateral aid. For instance, Kaja and Werker (2010) show that developing countries serving on the World Bank's Board of Executive Directors tend to receive twice as much aid as non-board members from the Bank's two main development financing institutions, the International Bank for Reconstruction and Development (IBRD) and the International Development Association (IDA). Their findings highlight how internal politics on the Executive Board can shape the allocation of multilateral development assistance.

[7] For example, in their study of "who gives aid and to whom," Alesina and Dollar (2000) show that Scandinavian donors are less geopolitically motivated in their aid allocation compared to major powers.

It is worth noting that development assistance may bypass either a donor or recipient *government* (or both). For instance, nongovernmental organizations (NGOs) such as the Gates Foundation can disburse funds to recipient governments, as well as to NGOs in recipient countries. The latter type of recipient highlights that aid can "bypass" governments.[8] While these alternative modalities are increasingly important, they are substantially smaller (in absolute dollars) than aid involving governments, with the United States as the world's largest bilateral donor. The political economy of US aid is discussed in greater detail in the next few subsections.

Importance of US Aid

Since 1960, the United States has allocated more than $700 billion in bilateral economic assistance, an amount exceeding that of any other nation. This figure excludes US military aid, aid to multilateral organizations (e.g., World Bank, regional development banks), and food aid. Figure 4.1 captures the temporal variation in total US bilateral economic aid (left scale) and as a share of total Development Assistance Committee (DAC) aid (right scale) since 1960. US economic aid averaged nearly $17 billion per annum in the 1960s, during a period of robust domestic economic growth coupled with relatively intense Cold War tensions. As superpower rivalry eased during the period of détente and the US experienced economic recession in the 1970s, US economic assistance fell to around $10 billion per annum (and maintained that annual average throughout the 1980s and 1990s). In the 2000s, primarily in response to the events of 9/11, US economic aid increased substantially. Since 2001, US economic aid has averaged more than $21 billion per year. The US share of total DAC aid has also varied over time, ranging from a high of 50 percent in 1963 to a low of 12 percent in 1997. Since 1960, aggregate US aid has amounted to 28.5 percent of total DAC bilateral assistance, which exceeds the share of all other bilateral aid donors.[9]

The Politics of US Aid

Since World War II, the US government has disbursed aid as part of its broader foreign policy objectives.[10] In official statements, the US government declares that its aid aims to promote political liberalization: "The protection of fundamental human right was a foundation stone in the United States over

---

[8] Werker and Ahmed (2008); Dietrich (2016).
[9] Moreover, compared to the other four largest bilateral donors (France, Germany, Japan, and the United Kingdom), US aid also tends to be more volatile (annually). From an econometric standpoint, this greater variability is advantageous, as it will generate more precise estimates of the effect of US aid on political rights.
[10] Morganthau (1962); Baldwin (1986).

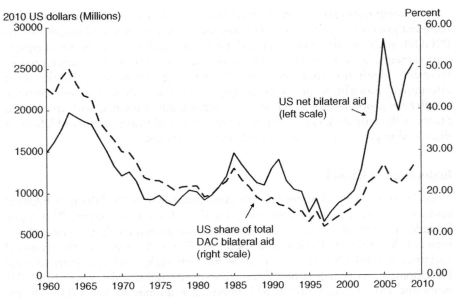

FIGURE 4.1 US bilateral economic aid, 1960–2009.
Source: Author's calculation based on data from the World Development Indicators.

200 years ago. Since then, a central goal of U.S. foreign aid has been the promotion for human rights, as embodied in the Universal Declaration of Human Rights."[11]

Scholars have investigated the veracity of these promulgations. Some evidence points to a positive association. Looking at a small subset of US economic aid, Finkel et al. (2007) find that "democracy aid" at "best" has modest effects on improving human rights.[12] Their finding corroborates Liang-Fenton's (2004) detailed qualitative analysis from fourteen country cases that US aid has yielded mixed effects on political liberalization, perhaps because of changing geopolitical priorities. For instance, since the end of the Cold War Western donors have increasingly channeled aid toward political liberalization

---

[11] Official statement on the US Department of State website, www.state.gov/j/drl/hr/index.htm
[12] Finkel et al. (2007, 410) argue aid can foster political liberalization in two ways: "indirectly, by transforming some of the structural conditions that serve as prerequisites for regime transition or survival, and directly by empowering agents (individuals, political institutions, and social organizations) that struggle for regime change in the domestic arena." Aid can also accelerate political liberalization for reform-minded governments. For instance, Wright (2009) shows that dictators with large distributional coalitions and who have a good chance of winning fair elections tend to respond to aid by democratizing.

4.1 Foreign Aid and Politics

in recipient countries.[13] These studies, however, do not establish a causal relationship.[14]

While the United States may allocate aid to help improve political rights in some contexts (e.g., induce political reform in post-Soviet states in Eastern Europe), in others it disburses aid to help prop up its autocratic allies. For example, US aid has played a prominent role in financing authoritarian rule in Egypt since the 1970s. As Arafat (2009, 164) declares, "For over decades, Egypt received the second-largest package of foreign aid from the United States (behind Israel), aid that many Egyptians believe views as supporting the Egypt's authoritarian regime, not its people." A by-product of this aid is the maintenance of peace between Israel and Egypt. More broadly, propping up dictators with foreign aid is often a cheap way for the United States to "buy" policy concessions from autocrats, where a smaller amount of funds is required to appease their smaller winning coalition, relative to the larger winning coalitions in democracies.[15]

Relatedly, the broader geopolitical objectives of the United States are an influential driver of its foreign aid decisions. This was apparent during the Cold War, where its geopolitical rivalry with the Soviet Union compelled it to buy allegiance from Third World dictators and sustain their repressive rule. Foreign assistance to Somalia during the Cold War is a prime example of how US economic assistance – in conjunction with aid from US allies – financed the repressive dictatorial rule of General Siad Barre.

Case Vignette: Aiding Repression in Somalia

Barre's Repression. The relationship between aid and state repression is starkly apparently in the case of General Siad Barre's Somalia. After gaining its independence in 1960, Somalia embarked on forging a parliamentary democracy. This government was short lived, as army officer Siad Barre led a successful coup in 1969 that established a Siad-led Supreme Revolutionary Council as the ruling body. Adopting a self-styled "scientific socialist" framework, Barre initially forged ties with the Soviet Union, which provided

---

[13] See Dunning (2004), Finkel et al. (2007), and Bermeo (2011). It is worth noting, however, that geostrategic considerations may undermine the purported democracy enhancing intent of US aid. For instance, aid can advance geopolitical objectives by potentially "buying support" in the UN General Assembly (Lai and Morey 2006), as well serving as a "signal" to attract foreign direct investment into developing countries (Garriga and Phillips 2014). Relatedly, Boschini and Olofsgard (2007) find that the end of the Cold War contributed to a significant reduction in US aid disbursements, while in postconflict environments, aid is effective in *less* geostrategic settings (Girod 2011). "Geostrategic considerations" are explicitly accounted for in this chapter's cross-national empirical analysis by controlling for country and year fixed effects, membership on the UN Security Council, and trade and alliance ties with the United States.

[14] See Ahmed (2016) for a detailed exposition of how many extant studies (as of 2018) do not provide valid causal estimates of the effect of US aid on political liberalization.

[15] Bueno de Mesquita and Smith (2009).

his nascent regime with substantial aid through the mid-1970s.[16] However, Barre's irredentist drive to reclaim Ogaaden Somali territory in Ethiopia 1977 fractured relations with the Soviet Union (since the USSR also provided assistance to Ethiopia). As a consequence, the Soviet Union withdrew its support for Barre, providing an opportunity for the United States – in the context of Cold War geopolitics – to gain Somalia's allegiance in a strategically important region of the world.

After switching to the US side, Barre governed with increasingly repressive and ruthless tactics based on clanism. Like in most African countries, group affiliations based on ethnic or tribal kinship forms the basis of political organization and consolidation. In particular, clanism is the Somali version of ethnicity and tribalism: it represents primordial cleavages and cultural fragmentation within Somali society.[17] While the postindependence parliamentary regime (1960–1969) sought to shift away from clan-based political organization in favor of an emerging commercial and bureaucratic petty bourgeoisie, Barre rejected such a strategy. Rather, on Barre's coming into power in 1969, his military regime centralized power via coercion: "the regime used a narrow clan base while condemning and denying the political space for other clan bases."[18]

Like many other African leaders, Barre instituted personal rule, evolving from being an advocate of "scientific socialism" (1970–1977), to an autocrat (1978–1986) and finally a tyrant (1987–1991).[19] Barre's abandonment of scientific socialism in favor of autocratic rule (beginning in 1978) coincided with his switch to the US side in the Cold War. In rejecting scientific socialism, Barre introduced "clan-klatura" as a governing ideology, in which "foreign aid provided the glue that held the system together in spite of internal waste and corruption."[20] Clan-klatura involved placing trusted clansmen and other loyalists in positions of power, wealth, and control.[21] After the Ogaden War (1977–1978), Barre initiated a brutal divide-and-rule strategy that encouraged outright clan warfare. Adam (1999, 174) summarizes the strategy: "At first he used his army to conduct punitive raids, similar to those under early colonial

---

[16] Besteman (1996).

[17] As Adam (1999, 170) notes: "After the Second World War, clannism among Somalis favored nationalism and a Greater Somalia. At times, however, it has assumed a negative aspect – the abandonment of objectivity when clan and local/parochial interests have prevailed. Clan consciousness is a partly a product of elite manipulation – the cooptation and corruption of politicians claiming clan leadership – but at times it is the elite itself, that is manipulated by politicized clannism."

[18] Adam (1999), 171.   [19] Jackson and Rosberg (1982).   [20] Adam (1999), 173.

[21] See Lewis (1988). Barre's regime favored individuals from three clans from the Darod clan-family, given the clandestine code name "M.O.D." As Lewis (1988, 222) notes, "M (Marehan) stood for the patrilineage of the President, O (Ogaden) for that of his mother, and D (Dulbahante) for that of his principal son-in-law, head of the National Security Service. . . . [Though] no could utter the secret symbol of General Siyad's power openly, the M.O.D. basis of his rule was public knowledge and discussed and criticized in private."

4.1 Foreign Aid and Politics

TABLE 4.1 *Political institutions and political repression in Somalia, 1960–2010*

Decade	POLITY2	Executive constraints	Political rights	Civil liberties	Political terror
1960s	5.6	6.4			
1970s	–7	1	7	6.42	2.75
1980s	–7	1	7	7	3.6
1990s	–0.7	1	7	7	4.5
2000s	0	n.a.	7	7	4.3

Notes: Decade averages for each index. POLITY2 ranges from –10 (Least democratic) to +10 (Most democratic). Executive constraints ranges from 1 (Least constrained) to 7 (Most constrained). Political rights and civil liberties range from 1 (Most free) to 7 (Least free). Freedom House categorizes scores of 6 and 7 as "not free." The Political terror scale is based on reports from the US Department of State and ranges from 1 (Least terror) to 5 (Most terror).

rule. Later his troops armed so-called loyal clans and encouraged them to wage wars against 'rebel' clans. The damage caused by negative and destructive elite manipulation of clan consciousness contributed to the inability of civil society to rebound when Siyad fell from power."

The extent of Barre's ruthlessness is evident with various quantitative measures. Table 4.1 summarizes the average level of political repression since 1960 (by decade) across various indices, where increasingly higher values correspond to greater repression. The first decade after independence marked an era of democratic governance, where the average POLITY score fell just shy of a fully institutionalized democracy (POLITY score of +6 or greater) that characterizes all democracies. The ratification of a new constitution in 1961 delineated the powers of government, in particular placing considerable constraints on executive power. Based on the POLITY IV project's categorization of constraints on the chief executive, during the 1960s, the Somali prime minister faced "substantial" limitations on his authority that were on "parity" or almost "subordinate" with other political institutions (e.g., political parties, the legislature).[22]

The ascension of Siad Barre in 1969 bought an abrupt end to democracy. For the next 20 years, Somalia was autocratic with an average POLITY2 score of –7 where the leader enjoyed "unlimited" authority ($XCONST = 1$). Political freedoms were severely curtailed. Since 1970, Freedom House has rated Somalia as "not free" (at the maximal possible value of 7 on its political rights and civil liberties indices). The US Department of State measures of "political terror" continuously rose under Barre's rule from a limited amount

[22] During the 1960s, the average value of the executive constraints index in Somalia was 6.4, which lies in between "substantial limitations" on executive authority ($XCONST = 5$) and "executive parity or subordination" ($XCONST = 7$).

of imprisonment for nonviolent political activity (PTS = 2) in the 1970s to a greater incidence of executions and other political murders (PTS = 3) in the 1980s. Since the end of Barre's rule (ca. 1990) violations of civil and political rights had expanded to larger numbers of the population (PTS = 4) and teetered on terror throughout the whole population (PTS = 5).

Aiding Barre's Repression. The availability of external assistance proved critical in prolonging Barre's repressive regime.[23] Early in Barre's rule, the Soviet Union provided substantial military and economic assistance. After Barre's failed irredentist drive in 1977 and "switch" to the US side in the Cold War, the United States replaced the Soviet Union in providing economic and military assistance, providing about $100 million in economic aid per annum during the 1980s.

For the United States, Somalia's proximity to the Middle East and the Persian Gulf made it a strategic prize, especially in the changing geopolitical landscape. The late 1970s saw the demise of the Shah of Iran, the humiliating takeover of the US Embassy in Iran by revolutionaries, and the US perception of a growing Soviet presence in the Horn of Africa and Persian Gulf. In an effort to monitor affairs in the region, the United States doled out substantial financial assistance to Barre to secure military bases in Somalia. According to Rawson (1994), between 1980 and 1988, the United States provided $163.5 million in military technology and more than four times that amount in economic aid.[24] Figure 4.2 plots annual US economic aid disbursements to Somalia from 1960 until the end of Barre's rule. During the 1960s, US economic assistance averaged around $32 million per annum, but plummeted once Barre gained power in 1969 and subsequently the country's pledged allegiance to the Soviet Union. After switching to the US side in 1977, bilateral economic assistance from the United States increased, averaging around $92 million per annum from 1980 to 1990. For the entire period during the Cold War in which Barre "allied" with the United States (1977–1990), the United States allocated more than $1.1 billion in economic assistance to his regime.[25] This made the United States Barre's most generous bilateral patron. Cold War allies of the United States also provided aid to the Barre regime.[26] Throughout the 1980s, donor funding to Somalia soared, with contributions from Italy, the United Kingdom, Germany, and Saudi Arabia. Barre's regime also received financial assistance

---

[23] Adam (1999); Rawson (1996); Besteman (1996).
[24] As Besteman (1996) notes that in 1987, despite rampant human rights abuses and signs of imminent domestic unrest, the United States began building in Mogadishu one of its biggest embassy compounds in the world – including beachfront property, two swimming pools, tennis courts, and a golf course – at a cost of $35 million.
[25] The amount of bilateral financial assistance is much higher if US military and technical assistance is included.
[26] Donors, for example, invested in several projects, including a north–south tarmac road, a cigarette and match factory, a sports and theatre complex, and rice and tobacco farms. They also provided light arms and spare parts.

4.1 Foreign Aid and Politics

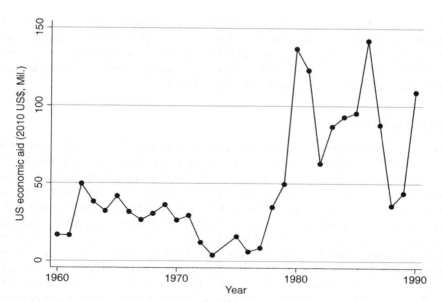

FIGURE 4.2 US economic aid to Somalia, 1960–1990.
Source: World Development Indicators. US economic aid to Somalia measured in millions of US 2010 dollars.

from the UN system and the World Bank. And after successfully maneuvering Somalia's entry into the Arab League in 1974, Barre garnered generous financial support from Arab oil producers.[27] In total, the Barre regime received at least $10.5 billion (2010 US$) in official development assistance (which excludes military assistance) from the United States and other donors. This aid helped finance Barre's repressive apparatus. According to Adam (1999, 177): "International actors helped to worsen state–society relationships in Somalia; their military, technical, and economic aid encouraged the dictatorship to believe that, because of foreign backing, it was capable of imposing its will on society. In spite of losing Soviet military and economic aid in 1977, Somalia continued to receive aid from the United States, Libya, France, Egypt, Italy, and China. The regime had reason to believe that, with such foreign backing, it could proceed against the opposition with impunity."

In the late 1980s, as United States strategic incentives changed and Barre's horrendous human rights policies gained greater public scrutiny (especially from the US Congress), economic assistance to Somalia tanked. Without external financing, Barre's grip on power waned and led to his destruction in the form of "civil wars of revenge" in which excluded clans rose up to challenge

---

[27] Neumayer (2003) and Ahmed and Werker (2015). Like other donors, some Arab countries used aid for strategic purposes.

the military dictatorship.[28] While the United States spearheaded a multilateral effort to provide humanitarian relief through the United Nations, the intervention was short lived and did not end the fighting. Since 1990, there have been more than 300,000 casualties and Somalia is presently a fragile state plagued with ineffective governance and violence.

Endogeneity Bias

While Barre's Somalia illustrates how a dictator can use foreign aid – especially assistance from the United States – to repress his population, one case is insufficient to demonstrate a robust, causal effect. Yet establishing causality in a cross-national setting is challenging, owing in large part to endogeneity bias. For instance, untangling the causal impact of US aid on measures of repression is problematic because US disbursements are frequently correlated with a recipient's political conditions. On the one hand, US aid may reward countries committed to political liberalization, such as assistance to help nascent Eastern Europe transition to democracy after the Cold War. On the other hand, US aid may help stabilize autocratic allies (e.g., Egypt) and thus undermine political liberalization.[29]

To overcome this challenge, I leverage an instrumental variables strategy that exploits two sources of variation. The first builds on the institutional foundations of US aid decisions, in which the funding and allocation of bilateral economic aid involve both the executive and legislative branches of government. Congress, in particular, is legally responsible for determining the aid budget. The second utilizes variation in the frequency of US aid distribution: countries that receive aid more frequently tend to receive higher amounts of aid and with less variation from year to year. As discussed in the next section, these two sources of variation form the basis of an instrumental variable for US bilateral aid.

4.2 IDENTIFICATION STRATEGY

Legislative Determinants of US Aid Allocation

Legislative Fragmentation. The United States allocates varying amounts of bilateral economic aid to recipient countries over time. A large component of this allocation process is influenced by US domestic politics. The majority of US foreign assistance is contained in the international affairs budget requested and allocated through the State, Foreign Operations, and Related Agencies appropriations bill in the US Congress. The legislative branch plays a critical role in US foreign assistance, possessing the power both to authorize policy and

---

[28] Laitin (1999).
[29] Indeed, most extant studies that attempt to gauge the effect of aid – in particular of US bilateral assistance – on political liberalization tend to sidestep these concerns with endogeneity. See Ahmed (2016) for a discussion of the limitations of existing research. Consequently, these studies do not causally identify the impact of US foreign aid on political rights.

4.2 Identification Strategy

appropriate funds. In response to the president's budget submission (by February 2 every year), the House and Senate budget committees are the first to act, setting funding ceilings for various parts of the budget and guiding the work of both authorizing and appropriations committees. Each year, eleven or twelve appropriations bills, including the State, Foreign Operations, and Related Agencies bill, make their way through a long deliberative process in both the House and the Senate. The appropriations committees, in coordination with the authorizing committees, determine and allocate federal spending each year, including foreign aid. Frequently, the resulting appropriations bills and accompanying reports include numerous detailed directives on how funds should be spent by country and account.[30]

This legislative process frequently reflects the interests of Congress.[31] Milner and Tingley (2010), for example, analyze votes related to US foreign aid from members of the House of Representatives from 1979 to 2003 and find that members with a more right-leaning political ideology tend to oppose economic aid than do members from more left-leaning districts.[32] Partisan affiliation often shapes the types of aid Congressmen support. For instance, analyzing US bilateral aid for 119 countries from 1960 to 1997, Fleck and Kilby (2006) show that a more liberal Congress (i.e., higher share of Democratic legislators to Republican legislators) gives greater weight to aid for economic development. In contrast, a more conservative Congress gives more weight to aid for commercial purposes (e.g., aid that is tied to US exports).

The existence of these partisan differences suggests that the legislative composition of Congress influences aid disbursements. In particular, existing theories and empirical evidence suggest that a more fragmented legislature contributes to higher government spending, including foreign aid appropriations.[33] The theoretical explanations stem from the well-established proposition that higher levels of aggregate political conflict (e.g., stemming from greater ideological/partisan differences in legislatures) will result in equilibrium fiscal outcomes that favor greater spending because politicians will exhibit a greater proclivity in providing voters with program benefits.[34] Moreover, greater heterogeneity in partisan preferences over fiscal policy is likely to require legislative logrolling, thus contributing to higher overall spending to accommodate different spending initiatives and to better ensure the bill's passage in Congress. A number of studies confirm this legislative fragmentation–spending relationship in the United States and elsewhere.[35]

---

[30] Lancaster (2000).   [31] Therein and Noel (2000); Milner and Tingley (2010).
[32] In contrast, House members from more right-leaning districts favor *military aid* than do members from less right-leaning district.
[33] Roubini and Sachs (1989); Alesina and Tabellini (1990); Alesina and Perotti (1996).
[34] Roubini and Sachs (1989); Alesina and Tabellini (1990).
[35] On cross-national evidence see Roubini and Sachs (1989) and Alesina and Tabellini (1990). For evidence from presidential systems, such as the United States, see Alesina and Rosenthal (1995) and Hankla (2013).

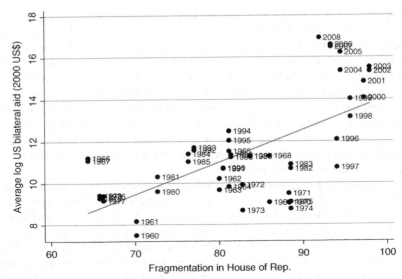

FIGURE 4.3 Fragmentation in the US House of Representatives and average US bilateral aid disbursements.

Such a relationship exists between the legislative fragmentation of the US House of Representatives and US bilateral aid disbursements. Figure 4.3 depicts a robust positive correlation between average US bilateral aid disbursements and a measure of legislative fragmentation based on the difference in the number of Democrats and Republicans in the House. Specifically, fragmentation ($FRAG_t$) in year t is defined as $\left(1 - \frac{|DEMOCRAT_t - REPUBLICAN_t|}{435}\right) \times 100$, where a higher value corresponds to greater fragmentation. Using the *absolute* difference in the number of House Democrats and Republicans avoids explicitly incorporating measures of partisanship or ideology (e.g., DW-NOMINATE), which are potentially endogenous with actual preferences for foreign aid.[36]

Exogeneity. Exploiting the legislative fragmentation from the US House (rather than from the US Senate) is advantageous for a number of reasons. First, all 435 members of House are subject to reelection every two years as opposed to only one-third of the 100 senators. Empirically, this means the House $FRAG_t$ exhibits greater temporal variation than the Senate $FRAG_t$ and generates a statistically stronger and more precise instrumental variable for US aid.[37] Second, and most importantly, $FRAG_t$ is a plausibly exogenous source of temporal variation in US aid disbursements that is *uncorrelated* with political

---

[36] Fleck and Kilby (2006); Milner and Tingley (2010).
[37] Aid instrumented using Senate fragmentation also hurts political rights. These results are reported in Ahmed (2016), appendix table B3.

4.2 Identification Strategy

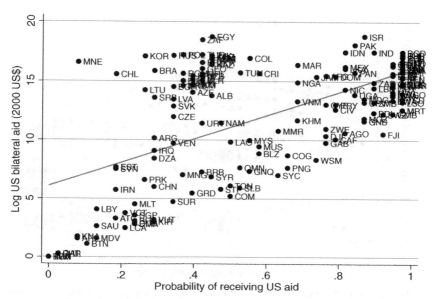

FIGURE 4.4 Annual probability of receiving US aid (P_i) and average US bilateral economic aid.

and economic conditions *within* aid recipients. Changes in the composition of the US House of Representatives occur biannually as a consequence of elections that are largely determined by local and national political and economic conditions, including (but not limited to) federal spending in Congressional districts, presidential coattails, midterm elections, and retrospective economic voting.[38] To the best of my knowledge, political conditions in poor developing countries have not been identified as a determinant in US House elections.

Aid Frequency. The United States tends to dole out higher amounts of aid to more frequent aid recipients. Figure 4.4 plots a country's average receipts of US aid (over the period 1972–2008) against its annual probability of receiving any US aid, P_i.[39] For instance, Nigeria has a 68 probability of receiving US aid in any given year, with aid disbursements averaging $31.3 million per annum. In contrast, Algeria receives a substantially lower amount of aid ($41,803 on average per annum) about once every three years. The cross-sectional relationship identified in Figure 4.4 is analogous to Nunn and Qian's (2014) observation that US bilateral food aid is higher for countries that receive food aid more frequently from the United States.

---

[38] On Congressional districts, see Levitt and Synder (1997); presidential coattails, see Campbell and Sumners (1990); midterm elections, see Tufte (1975); and retrospective economic voting, see Fiorina (1978).

[39] P_i is based on the proportion of years between 1972 and 2008 a country receives any US aid.

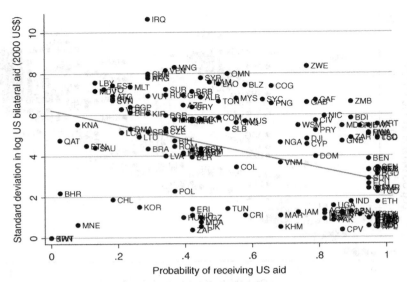

FIGURE 4.5 Annual probability of receiving US aid (P_i) and annual variation (standard deviation) in US bilateral economic aid (by country).

Variation in Aid Levels. More frequent aid recipients tend to have more stable aid receipts. Figure 4.5 shows that more frequent US aid recipients are *less* likely to experience *changes* in their annual aid receipts. It demonstrates this by plotting each country's probability of receiving US aid (P_i) against the standard deviation of its annual *level* of US aid inflows. The negative association implies that *less* (*more*) *frequent aid* recipients exhibit *greater* (*less*) *variation* in annual aid disbursements. Together the relationships in Figures 4.4 and 4.5 imply that *more* (*less*) frequent aid recipients are *less* (*more*) sensitive to "shocks" in the *total* aid budget stemming from changes in the composition of Congress. Empirically, this means the *interaction* of $FRAG_t$ and P_i will be *negatively* associated with US aid receipts, an effect that the first-stage regression demonstrates (see Table 4.2, column 1).

Instrumental Variable

I exploit these sources of variation to construct a powerful cross-national and time-varying instrumental variable for US aid. The instrument interacts the legislative fragmentation of the US House of Representatives ($FRAG_t$) with the probability a country receives US aid in any year (P_i). This instrument ($FRAG_t \times P_i$), which is constructed by interacting a plausibly exogenous term ($FRAG_t$) with one that is potentially endogenous (P_i) can be interpreted as

4.2 Identification Strategy

exogenous because the first-stage and second-stage regressions control for main effect of the endogenous variable (see equations that follow).[40]

Armed with this instrumental variable, the two-stage least squares (2SLS) setup is

First stage : $AID_{it} = \alpha + \beta (FRAG_t^* P_i) + \gamma X_{it} + Y_t + C_i + \varepsilon_{it}$
Second stage : $RIGHTS_{it} = a + b^* AID_{it} + c^* X_{it} + Y_t + C_i + u_{it}$

where i refers to the country, t indexes the year, X_{it} is a vector of controls, and C_i and Y_t are fixed effects for countries and years, respectively. The inclusion of country fixed effects implies the estimated coefficients will gauge each aid recipient's within-country variation in $RIGHTS_{it}$. To account for potential serial correlation, the standard errors are conservatively clustered by country.

In the first-stage regression, AID_{it} is each country's annual receipts of US bilateral economic aid. As Scott and Steele (2011) argue, because aid appropriations and obligations can go unspent, I use disbursements to evaluate the impact of "realized aid."[41] Following several prominent studies, aid is measured in logarithmic units (i.e., $\log(1 + AID_{it})$).[42] In the second-stage regression, $RIGHTS_{it}$, is an increasing measure of political rights, where higher values imply *greater* political repression. This means that if the United States harms political rights, the coefficient on AID_{it} (b) should be positive and statistically significant.

---

[40] The identifying assumption is that the "endogenous" variable and the outcome of interest are jointly independent of the "exogenous" variable. See section 2.3.4 of Angrist and Krueger (1999) for a more technical discussion. Constructing an instrumental variable for aid in this fashion underlies the identification strategy of several prominent articles in the foreign aid literature. For example, in their study of the impact of US food aid on civil conflict, Nunn and Qian (2014) interact plausibly exogenous variation in annual US weather conditions with the probability a country receives US food in any particular year as an instrumental variable for US food aid allocations. The latter term in their instrumental variable (probability of receiving US food aid) is country specific, time invariant, and *identical* to the formulation of P_i employed in this chapter. Among the exhaustive robustness checks conducted by Ahmed (2016), the results also hold in specifications that do not include fixed effects, but do include the relevant constituent terms of the instrumental variable (e.g., Table 4.2, column 2). The core results also hold in specifications with time-varying measures of P_i and alternate derivations of P_i based on the aid allocation decisions of the British and Canadian governments.

[41] Other related studies also use disbursements, such as Wright (2009), Steele and Scott (2011), Nunn and Qian (2014), and Ahmed and Werker (2015).

[42] Some of these prominent studies include Alesina and Dollar (2000), Kuziemko and Werker (2006), and Boschini and Olofgard (2007). There are both theoretical and empirical reasons for measuring total aid (in log units). Several formal models link a country's *level of total receipts* to repression (e.g., Besley and Persson 2011), rather than per capita aid or aid as a share of GDP. Empirically, many scholars note that log total aid reduces skewness in the underlying measure of aid flows and accounts for potentially diminishing returns to aid (on the relevant outcome variable). Finally, several prominent studies in the empirical literature use log total aid as their independent variable (Kuzeimko and Werker 2006), as well as their dependent variable of interest (e.g., Boschini and Ologsfgard 2007).

Turning to the instrument, $FRAG_t$ is equal to $\left(1 - \frac{|DEMOCRAT_t - REPUBLICAN_t|}{435}\right) \times 100$, where a higher value (i.e., closer to 100) implies greater legislative fragmentation in the House of Representatives. The tendency for a country (i) to receive any aid is given by $P_i = \frac{1}{38} \sum_{t=1972}^{2009} P_{i,t}$, where $P_{i,t}$ is equal to 1 if that country receives any aid in year t and zero otherwise. Observe that the vector of country fixed effects absorbs this probability because P_i is specific to each country (i) and time invariant. The inclusion of year fixed effects subsumes the main effect corresponding to legislative fragmentation because $FRAG_t$ changes from year to year, but remains the same across all aid recipients. Year fixed effects also account for any constant time trend in the independent and dependent variables.[43]

In both stages, I control for a parsimonious set of covariates (X_{it}) that affects both the allocation of aid in the first-stage regression and $RIGHTS_{it}$ in the second-stage regression: log GDP per capita ("need") and economic growth ("merit").[44] I also control for a country's population size because smaller countries tend to receive disproportionately higher amounts of aid, and the cost of political repression often varies by country size.[45] These control variables also serve to account for the main channels through which "modernization theories" can foster political liberalization.[46]

Measuring donor self-interest, in contrast, is not as obvious as their motives are numerous, often donor specific, and largely unobservable. The inclusion of country and year fixed effects, respectively, will account for time-invariant characteristics (e.g., colonial ties) and "global" factors (e.g., Cold War politics, global terrorism) that might affect a donor's interest in allocating aid. Geopolitical concerns, of course, are also often time varying and country specific. For instance, Kuziemko and Werker (2006) show that countries that are elected to the United Nations Security Council (UNSC) receive higher amounts of US aid, some of which can lead to pernicious political and

---

[43] For instance, Figures 4.1 and 4.3 suggest that the overall US aid budget (Figure 4.1) and average disbursements to countries (Figure 4.3) have tended to increase over time. The inclusion of year fixed effects accounts for this upward trend. The main results in this chapter also hold in specifications with a time trend. It is worth emphasizing that the chapter's results do not hinge on the inclusion of fixed effects or on the construction of the instrument in this particular way. The findings are robust across a variety of different regression specifications and formulations of the instrument. For example, in first-stage regression specifications that include the constituent terms of the instrumental variable (i.e., $FRAG_t$, P_i) as control variables (Table 4.2, column 1), the coefficients on $FRAG_t$ and P_i are both positive and significant (in line with the positive trends in Figures 4.3 and 4.4), while their interaction ($FRAG_t \times P_i$) exhibits a negative relationship with aid (in line with the negative trend in Figure 4.5).

[44] According to Hoeffler and Outram (2011), foreign aid is often channeled to poorer (i.e., "needier") countries; moreover, donors often reward those poor countries that are exhibiting signs of economic growth (i.e., "merit").

[45] Alesina and Dollar (2000).   [46] As argued by Finkel et al. (2007), for example.

economic outcomes.[47] Relatedly, US allies receive higher amounts of aid. Finally, US aid is often aimed at promoting US exports to aid recipients and greater import penetration has been linked to better governance in countries.[48] To account for these effects, I control for a recipient's annual consumption of US exports, its membership in the UNSC, and its alliance status with the United States.

Data

Dependent Variable. I use Freedom House's POLITICAL RIGHTS index as the main measure of political repression. This index has been used in similar studies and has the largest country (~150) and temporal coverage (1973 onwards) compared to related measures, such as those from POLITY and CIRI Human Rights Project.[49] Based on the opinions of experts, POLITICAL RIGHTS measures the ability of "people to participate freely in the political process, which is the system by which the polity chooses authoritative policymakers and attempts to make binding decisions affecting the national, regional, or local community" (e.g., the right to vote, the capacity of elected officials to have decisive votes on public policies). The index lies on a 1- to 7-point scale, where *higher* values (e.g., 6 or 7) correspond to *less* freedom.

Independent Variables. The key independent variable, AID_{it}, is the United States' net disbursements of official development assistance (ODA). There is wide cross-national and temporal variation in US economic aid. Some countries (e.g., Bhutan, Maldives) receive very little US aid (i.e., less than $1 million), while some countries receive aid exceeding $10 million per annum on average (e.g., Bangladesh, El Salvador) and several surpassing $500 million annually (e.g., Egypt, Israel, Iraq after 2003). In the baseline sample, the typical country receives about $50 million in US economic aid per annum. The economic and demographic controls are drawn from the World Development Indicators.[50] UN Security Council membership is available from the United Nations, alliances from the Correlates of War, and US exports from the International Monetary Fund.[51]

4.3 US AID REPRESSES

US Aid Harms Political Rights

Determinants of US Bilateral Aid. Table 4.2 shows that the instrumental variable ($FRAG_t \times P_i$) is a robust determinant of US bilateral aid to about

---

[47] On these pernicious effects, see Bueno de Mesquita and Smith (2010b).
[48] Fleck and Kilby (2006) evaluate how US aid affects the purchases of US exports by aid recipients. On the relationship between imports and governance, see Ades and di Tella (1999).
[49] Finkel et al. (2007); Kersting and Kilby (2014). As robustness, I also show that US aid hurts political rights using measures from CIRI.
[50] World Bank (2010).   [51] Gibler and Sturkes (2004); International Monetary Fund (2012).

TABLE 4.2 US aid harms political rights

	First-stage regression			Second-stage regression		
Dependent variable:	Economic aid	Economic aid	Economic and military aid	Political rights		
	(1)	(2)	(3)	(4)	(5)	(6)
$FRAG_t \times P_i$	−0.343	−0.342	−0.314			
	(0.068)***	(0.068)***	(0.062)***			
Economic aid				0.149	0.157	
				(0.069)**	(0.059)***	
Economic military aid						0.171
						(0.064)***
Economic aid in $t-1$						
Fragmentation ($FRAG_t$)	0.37			−0.048		
	(0.052)***			(0.010)***		
Probability of receiving aid (P_i)	40.73			−2.13		
	(5.384)***			(1.012)**		
Country FE	N	Y	Y	N	Y	Y
Year FE	N	Y	Y	N	Y	Y
R^2	0.44	0.66	0.65	0.3	0.64	0.64
No. of observations	3,853	3,853	3,853	3,853	3,853	3,853
F-statistic	35.14	25.32	25.39			

Notes: Estimation via ordinary least squares. Robust standard errors clustered by country are reported in parentheses. *, **, *** = statistically significant at 10%, 5%, and 1% respectively. Economic and military aid is measured in log units (2000 US$). All specifications control for the following recipient characteristics: log GDP per capita (2000 US$), GDP per capita growth (% annual), log population, UN Security Council member, US ally, and log US exports (2000 US$). These coefficients, a constant, and country and year fixed effects are not reported.

150 countries. Column 1 presents a specification without any fixed effects. Consistent with Figure 4.3, greater legislative fragmentation raises US aid disbursements (coefficient = 0.37). And consistent with Figure 4.4, more frequent aid recipients receive higher amounts of aid (coefficient = 40.73). The coefficient on the instrumental variable is −0.34 and is highly statistically significant ($p = 0.00$). This negative coefficient is consistent with Figure 4.5, in which *more* frequent aid recipients tend to experience *less variation* in annual US aid receipts. The corresponding F-statistic (= 35.14) means the instrument is "strong" (because it exceeds the threshold of 9.6 suggested by Stock et al. 2002) and implies the second-stage estimates can be interpreted as causal. Column 2 shows the instrument to be a strong predictor of US

4.3 US Aid Represses

economic aid in a specification with country and year fixed effects.[52] Column 3 shows that the instrumental variable is also a strong predictor of US economic *and* military aid.[53]

The Impact of US Aid on Political Rights. Columns 4 to 6 in Table 4.2 report the effect of instrumented US aid on political rights. Column 4 shows that US aid *causes* a deterioration of political rights: a unit increase in log US economic aid raises POLITICAL RIGHTS by 0.15 index point. This effect is highly statistically significant ($p < 0.01$) and substantively meaningful: moving from the 10th percentile of aid receipts to merely the 50th percentile corresponds to around a 2.5 index point rise in POLITICAL RIGHTS. Such a jump corresponds to a 1.25 standard deviation increase in the POLITCAL RIGHTS index and is equivalent to moving from a less repressive country such as Peru to a more repressive country such as Sudan or Vietnam.[54] The control variables (not reported) are consistent with existing findings. For example, richer countries (i.e., higher GDP per capita) are less repressive. The inclusion of country and year fixed effects does not attenuate the effect of US economic aid on political effects (column 2). Rather, the estimated effect is slightly larger (= 0.16). Instrumented US economic and military aid also harms political rights (column 3).

Robustness. The core result in Table 4.2 is robust to an exhaustive list of other potential concerns, such as outliers (e.g., exclusion of frequent aid recipients); spatial "diffusion"; the inclusion of additional controls (e.g., political institutions, oil exports, arms imports, overall trade "openness," conflict, percentage of born population); alternate instruments (e.g., using fragmentation in the Senate); alternate specifications (e.g., controlling for lags and leads of aid, time-varying measures of P_i); alternate clustering of the errors (e.g., by region, year, two-way clustering); the "crowding-out" of aid from other donors; regional trends (e.g., differential effects for Africa); and region- and country-specific trends that vary across the Cold War and post–Cold War period.[55]

Considerations of the importance of geopolitics are particularly noteworthy, as recent research suggests that after the Cold War, Western donors have increasingly channeled aid to recipients that are committed to improving democracy.[56] Whether such aid causes improvements in democracy has *not*

---

[52] This specification excludes $FRAG_t$ and P_i because they are subsumed by the relevant set of fixed effects (year and country, respectively).

[53] In the first-stage regressions, the control variables have their expected effects (not reported). In general, richer countries receive lower amounts of aid, while high-growth countries are "rewarded" with more aid. The time-varying geostrategic measures to not seem to affect US aid. These results are reported in Ahmed (2016), table 1.

[54] Based on average POLITICAL RIGHTS for each of these countries over the sample period.

[55] These robustness checks are reported and discussed in Ahmed (2016).

[56] Bermeo (2011); Dietrich and Wright (2015). The data suggest this correlation, as the United States tends to dole out more aid to democratic recipients (i.e., P_i tends to be higher for countries with higher POLITY scores).

been established in the literature, especially after the Cold War. In a sample of country–year observations that is restricted to the post–Cold War period only (1990 onwards), instrumented aid does not improve POLITICAL RIGHTS. Rather, in such a model, the effect on instrumented US aid on POLITICAL RIGHTS remains positive and statistically significant (coefficient = 0.081, standard error = 0.041, $p < 0.05$). Relatedly, skeptics may worry that the surge in US aid after the 9/11 terrorist attacks in 2001 (see Figure 4.1) may be unduly influencing the findings. To mitigate this worry, I evaluate the effect of US aid on POLITICAL RIGHTS for a sample of country–year observations *prior* to 2001. With this pre-2001 sample, the effect on instrumented US aid on POLITICAL RIGHTS remains robust and the estimated effect actually increases in magnitude (coefficient = 0.214, standard error = 0.126, $p < 0.10$).

Finally, the core finding in Table 4.2 is not sensitive to how I measure political rights. US aid also harms other forms of human rights (and from different data sources): civil liberties, political participation, and freedoms associated with religious and empowerment rights and imprisonment.[57] These robustness checks are reported and discussed in Ahmed (2016).[58]

4.4 EVALUATING CHANNELS

Repressive Capacity versus Tax Effort

How does US aid harm political rights? Chapter 3 argues that repression can occur through two plausible channels. One channel posits that foreign aid can fund private government consumption, such as expenditures on increasing the state's repressive capacity (e.g., strengthening the military). The second channel posits that foreign aid can sustain a government's *existing* expenditures while *also* allowing it to reduce its tax effort. By relieving this need to tax, foreign aid lowers the government's incentive to be accountable to its population. Moreover, these channels are not necessarily mutually exclusive: a government may respond to additional aid by increasing its repressive capacity *and* by decreasing its tax effort. Panels A and B in Table 4.3 evaluate these channels.

Panel A in Table 4.3 provides evidence that US aid may shift the composition of government expenditures in recipient states. The direction of the coefficient estimates in columns 1a and 2a suggests that US aid shifts government spending away (i.e., a negative coefficient) from welfare goods provision in the form of payments on transfers and subsidies (column 1a) to the military (column 2a). In these regressions, the dependent variables are measured as a *share* of total government expenditures; thus, the positive coefficient in column 2a implies

---

[57] These data sources are Freedom House (2011), Marshall and Jaggers (2010), and Cingaranelli and Richards (2008).
[58] These robustness checks are available as part of the book's replication materials.

4.4 Evaluating Channels

TABLE 4.3 *Evaluating the channels*

Dependent variable:	A: Military spending				
	Transfers	Military	Log military expenditures (2000 US$)		
	(% government expenditure)			Autocracies	Democracies
	(1a)	(2a)	(3a)	(4a)	(5a)
Method of estimation:	2SLS	2SLS	2SLS	2SLS	2SLS
Log US aid	−0.661	0.867	0.107	0.141	0.050
	(1.080)	(1.039)	(0.099)	(0.179)	(0.074)
No. of observations	795	800	2,043	1,164	879
No. of countries	110	108	135	74	61
R^2	0.89	0.54	0.87	0.89	0.86

Dependent variable:	B: Taxation			
	Income tax (share of government revenue)			
	(1b)	(2b)	(3b)	(4b)
Method of estimation:	OLS	OLS	2SLS	2SLS
Log US aid		−0.324	−1.864	−1.938
		(0.200)*	(1.045)*	(1.023)**
Political rights	−0.200			−0.050
	(0.238)			(0.556)
No. of observations	1,380	925	925	915
No. of countries	136	113	113	111
R^2	0.88	0.83	0.7	0.69

Notes: Estimation via two-stage least squares (2SLS). Robust standard errors clustered by country are reported in parentheses. *, **, *** = statistically significant at 10%, 5%, and 1% respectively. US economic aid is measured in log units (2000 US$). Recipient characteristics include log GDP per capita, GDP per capita growth, log population, UNSC member, US ally, and US exports. These coefficients, country fixed and year fixed effects, and a constant are not reported.

that recipient governments exert greater "effort" on the military with additional US aid receipts. This enhanced military effort is unsurprisingly associated with an increase in the total level of military spending across all regime types (column 3a), but the effect is about three times larger in autocracies (column 4a) than in democracies (column 5a). This difference in coefficient size is consistent with the arguments and figures in Chapter 3 that autocracies tend to spend more on their militaries.

However, the results in panel A should be viewed as suggestive at best since the coefficient estimates are not statistically significant at conventional levels. In contrast, panel B provides more compelling evidence for the second channel linking foreign aid to repression via a taxation channel. This channel posits that as a source of nontax revenue, aid inflows should allow governments to reduce their tax effort and consequently permit it to become less accountable to its population (i.e., more repressive). Empirically, therefore, the amount of taxes collected from individuals (as a share of total government revenue) should be negatively correlated with aid inflows. To test this mechanism, I regress taxes collected from income, profits, and capital gains (% government revenue) on US aid plus the baseline controls.[59] A reduction in this dependent variable implies a reduced exertion of tax effort because a government is able to derive a larger share of its revenue from nontax sources.

Empirically, POLITICAL RIGHTS are negatively correlated with tax effort (column 1b). This affirms the underlying logic of a negative relationship between political repression and taxation. Turning to the conjectured channel, US aid seems to reduce tax effort. In an ordinary least squares model, US aid is negatively associated with tax effort (column 2b). In column 3b, instrumented US aid has a much larger and statistically significant effect on tax effort. A 1 standard deviation increase in US aid, for instance, lowers tax effort by 13 percentage points. Moreover, column 4b shows that controlling for repression does not attenuate the negative effect of instrumented US aid on tax effort. In fact, the statistical precision of the effect of aid on tax effort is strengthened. Instrumented aid also exhibits a negative effect on the level (in dollars and log units) of taxes collected (not reported). Together the results in columns 1b–4b suggest that US aid hurts political rights by reducing a government's tax effort.

4.5 AIDING AUTHORITARIAN SURVIVAL

The willingness for governments to deteriorate political rights (using US aid) is rational if doing so does not weaken a country's underlying political institutional foundations nor threaten the political survival of leaders. This seems to be the case. Tables 4.4 and 4.5 present evidence that US aid bolsters authoritarian institutions and leaders.

Authoritarian Institutions

I first evaluate the impact of US aid on political institutions using three well-known measures from the POLITY data set. The first is the executive constraints index (*XCONST*), which measures the extent of institutionalized constraints on the decision-making powers of chief executives, whether they are individuals or collectives. *XCONST* lies on a 1- to 7-point scale, where *lower*

---

[59] Data on tax collection for a large set of developing countries is only available from 1990 onwards.

4.5 Aiding Authoritarian Survival

TABLE 4.4 *US aid fosters authoritarian stability*

Dependent variable:	XCONST	POLITY2	DURABLE		
			All countries	Autocracy	Democracy
	(1)	(2)	(3)	(4)	(5)
US economic aid	−0.126	−0.503	1.833	2.627	1.236
	(0.073)*	(0.253)**	(0.908)**	(1.490)*	(0.554)**
No. of observations	3,225	3,341	3,376	2,050	1,326
R^2	0.66	0.68	0.56	0.39	0.8
No. of countries	129	129	130	74	56

Notes: Estimation via 2SLS. Robust standard errors clustered by country are reported in parentheses. *, **, *** = significant at 10%, 5%, and 1% respectively. US economic aid is measured in log units (2000 US$). Recipient characteristics include: log GDP per capita, GDP per capita growth, log population, UNSC member, US ally, and US exports. These coefficients, country fixed and year fixed effects, and a constant are not reported. In columns 4 and 5 the sample is restricted to autocracies and democracies respectively.

values correspond to a reduction in constraints (i.e., an aggrandizement of executive control) and usually imply a decline in a country's democratic governance. The second is the broader POLITY2 index (−10 to +10) where higher values correspond to higher quality democratic institutions. The third, *DURABLE*, is a measure of regime durability. If a country experiences at least a 3-point change in its POLITY score from the previous year, *DURABLE* is reset to zero. If there is no such change, *DURABLE* increases by one point.

Table 4.4 presents evidence that US aid strengthens authoritarian politics. In column 1, a 1 standard deviation increase in US aid lowers *XCONST* by 0.75 index point. This is equal to about a one-third standard deviation change in *XCONST*. This reduction in *XCONST* implies US aid *causes* dictators to expand their powers. Column 2 shows that US aid deteriorates the quality of democratic political institutions: a 1 standard deviation increase in US aid lowers POLITY by 3 index points (or roughly 0.43 standard deviation of POLITY). US aid also tends to make a country's existing political institutions less prone to rapid changes (column 3). Yet the effects are not uniform across regime type. In nondemocracies, the effect of US aid on *DURABLE* is more than twice as large as the effect in democracies (columns 4 and 5). This differential effect is consistent with the notion that repression can be a prudent strategy for regime stability in autocracies.[60]

---

[60] Smith (2008).

Political Survival

The willingness of governments to repress their populations with US aid is rational if such a strategy *does not* threaten the political survival of leaders, particularly in nondemocracies. Indeed, US aid may be especially attractive and beneficial for the survival prospects of dictators who rule in *less* politically consolidated autocracies. In these less repressive regimes, the marginal benefit of additional US aid should tend to lower the likelihood of leader termination. Table 4.5 evaluates these conjectures.

Using the same outcome variable from Chapter 2, I measure political survival using the incidence of leader exit ($EXIT_{it}$). $EXIT$ is a binary variable equal to 1 if a leader in country i exits political office in year t, and 0 if the leader survives in office. In econometric models with this outcome variable, a negative coefficient estimate implies that covariate lowers the prospect of leader exit. And in these models of leader survival, I follow convention by controlling for duration dependency with a quadratic polynomial of the number of days a leader has survived in office.[61] All the specifications control for time-varying recipient characteristics (e.g., GDP per capita, growth) and year and country fixed effects.

I first confirm the notion that political repression can be an effective strategy for political survival. Column 1 in Table 4.5 shows the relationship between POLITICAL RIGHTS and EXIT. The negative coefficient (=−0.22) on POLITICAL RIGHTS implies that leaders that govern in more repressive states are less likely to lose office. This finding supports the conjecture that underlies the analysis in this chapter: political repression can be an effective strategy for leaders to preserve their tenure in office. While this is reassuring for the theory, the core of this chapter is to evaluate the causal impact of US foreign aid on political repression, which if effective should *not weaken* the political survival of autocrats.[62] The remaining specifications in Table 4.5 evaluate this by gauging the effect of US aid on leader survival.

In an extension to the model in column 1, the specification in column 2 examines whether leaders in more frequent US aid recipients exhibit a lower propensity to lose office. While the effect is not statistically significant at conventional thresholds, the negative coefficient on P_i provides suggestive evidence that this seems to be the case. On the margins, leaders in the modal US aid recipient ($P_i = 0.75$) are about 2.6 percent less likely to lose office (which is equivalent to a 15 percent reduction in the baseline rate of leader exit).[63] With respect to actual US aid disbursements, columns 3 and 4 show that additional US aid *does not threaten* the political survival of autocrats. Across all

---

[61] Carter and Signorino (2010).
[62] Chapter 3 argues that repression is not an effective strategy for political survival of *democratic* leaders.
[63] In column 2, about 17 percent of leaders exit office in any given year. A 2.6 percent reduction is therefore equal to a 15 percent decline in this baseline rate (2.6/17 = 15 percent).

4.5 Aiding Authoritarian Survival

TABLE 4.5 *Political survival in US aid recipients*

Dependent variable:		*EXIT*		
		Autocracies	All	Less repressive
	(1)	(2)	(3)	(4)
Political rights	−0.216	−0.076		
	(0.109)**	(0.039)**		
Probability of receiving US aid (P_i)		−0.243		
		(0.298)		
Log US economic aid (2000 US$)			0.003	−0.024
			(0.006)	(0.075)
No. of observations	3711	3973	1810	726
R^2	0.95	0.83	0.87	0.81
No. of countries	133	149	75	54

Notes: Estimation via probit (columns 1 and 2) and 2SLS (columns 3 and 4). Robust standard errors clustered by country are reported in parentheses. *, **, *** = significant at 10%, 5%, and 1% respectively. US economic aid is measured in log units (2000 US$). All regressions control for a duration dependence polynomial (number of days in office and number of days in office squared), recipient characteristics, and year and country fixed effects. These coefficients and a constant are not reported. Recipient characteristics include log GDP per capita, GDP per capita growth, log population, UNSC member, US ally, and US exports. In column 2, country year fixed effects are not included since P^i is country-specific and time invariant. In columns 3 and 4 the sample is restricted to autocracies only. In column 4, the sample is restricted to less repressive autocracies. These are country–year observations where POLITICAL RIGHTS is less than +6.

autocracies, US aid tends to have no effect on leader exit, as the coefficient on US aid in column 3 is statistically indistinguishable from zero.

This effect, however, may mask the more pronounced effect of US aid on leader survival in less consolidated autocracies. In these countries, political institutions are less durable, political rights are in the intermediate range of repression, and governments are more prone to challenges for power.[64] In these countries, therefore, US aid may be more effective in lowering leader termination because it can "buy" *more* repression. Looking at a subsample of nondemocracies with levels of repression below the median value (i.e., POLITICAL RIGHTS with scores less than +6), column 4 provides some suggestive evidence. The negative coefficient on instrumented aid suggests that higher amounts of US aid tend to lower the prospects that a dictator will lose office in less repressive states. However, the effect is not significant at conventional levels, owing in large part to a small estimating sample, and should be interpreted as suggestive at best. Nevertheless,

---

[64] Svolik (2012).

the results in Table 4.3 show that repression can be an effective strategy for lowering a leader's prospect of exiting office and US aid does not threaten the political survival of these leaders.

4.6 CONCLUSION

For donor countries and organizations, the principal aim of foreign aid is to foster economic development in recipient countries. And among the donor community, achieving this objective often requires political liberalization in recipient countries. This perspective underlies the stated intentions of world's largest bilateral aid donor: the United States. As an instrument of American economic statecraft, the United States claims to use its bilateral economic aid to promote its national interest by expanding democracy and free markets, while improving the lives of citizens in developing countries. The evidence in this chapter casts serious doubt on this assertion.

In Somalia, US economic assistance provided the means for General Siad Barre's clan-based repressive rule for over two decades. More broadly, the chapter's cross-national evidence demonstrates that US aid *causes* a deterioration of political rights, expands the powers of government leaders, enhances the durability of authoritarian institutions, and does not threaten their tenure in power. US aid fosters authoritarian politics, mainly by lowering a government's tax effort in exchange for less democratic accountability. US aid is also associated with a shift in government expenditures toward the military. Taken together, these causal effects imply that US aid should be viewed as an important source of nontax revenue in many developing countries that unfortunately further strengthens authoritarian politics in countries with already weak democratic institutions. The findings imply that US aid exhibits an income effect whereby dictators spend some fraction of aid on strategies that strengthen authoritarian politics, such as increased military spending and particularly heightened political repression.

5

Remittances and Autocratic Power

Throughout history a strong executive with limited constraints on his or her political authority has been a prerequisite for dictatorship.[1] Of course, the ability of autocrats to enjoy these low constraints on their rule requires sufficient revenues to repress and buy loyalty via patronage.

This chapter presents cross-national evidence that remittances can finance authoritarian politics, by lowering the constraints dictators face and extending their time in power. Endogeneity, however, plagues efforts to test whether a causal relationship exists between remittance income and authoritarian politics.[2] To overcome this empirical challenge, I leverage a quasi-natural experiment of oil price–driven remittance flows emanating from the Persian Gulf to non–oil-producing Muslim countries in North Africa, the Middle East, and South Asia.

The identification strategy exploits two sources of variation. The first is *temporal* and is associated with plausibly exogenous movements in oil prices. During periods of high oil prices, Gulf oil producers tend to import labor from non–oil-producing Muslim countries (e.g., Jordan, Pakistan); and these migrants in turn remit large amounts of their earnings to their home countries (many of whom tend to be nondemocratic). The aggregate remittances for these labor-exporting countries are thus correlated with the world price of oil. And the world price of oil is plausibly exogenous from political and economic conditions in poor remittance-receiving countries. For example, the Arab oil embargo following the Yom Kippur War in 1973 generated a decade-long period of high oil prices. The second source of variation is *cross-sectional* and is related to a recipient country's distance to the Mecca. This distance is a proxy for the fixed cost of migration to the Persian Gulf. Countries closer to the Persian Gulf (e.g., Jordan) experienced higher remittance inflows than those farther away (e.g., Bangladesh).

Thus, I exploit the exogeneity of world oil prices interacted with a recipient country's distance to Mecca to construct a time-varying instrument for

---

[1] Wintrobe (1998); Bueno de Mesquita et al. (2003); Gledistich and Ward (1997).
[2] This stems primarily from two sources: potential measurement error and reverse causality. The "endogeneity problem" is discussed in Section 5.2.

remittance income at the country level. Using this instrument, I show that remittances cause autocratic leaders to face lower constraints on their power, extend their duration in office, and make their country's political institutions less democratic. I also present evidence in support of the underlying channel: remittances reduce a government's expenditures on welfare goods in favor of increased spending on public sector compensation and corruption (both constitute patronage in autocracies).

This empirical finding that remittances can be a source of authoritarian politics offers a new perspective to understand the divergent political dynamics of Egypt and Jordan during the 2011 Arab Spring. In a comparative vignette, I argue that buoyant amounts of remittances provided sufficient external rents to sustain patronage in Jordan and hinder the ousting of the ruling monarch during the Arab Spring. This was not the case in Egypt, in which declining remittance income may have weakened the nondemocratic regime's ability to fund its various patronage networks (e.g., the military) and contributed to President Mubarak's removal from power.

I proceed as follows. Section 5.1 discusses the importance of migrant remittances and the potential channels through which they can affect politics in recipient countries, including an expenditure-switching mechanism (substitution effect). Section 5.2 describes the research design and data to evaluate the causal effect of remittances on authoritarian politics. The section introduces the quasi-natural experiment that forms the basis of the instrumental variables research design. Sections 5.3 to 5.6 present cross-national evidence that remittances entrench authoritarian politics via an expenditure-switching mechanism. I also present evidence to discount other channels through which remittances might affect authoritarian politics, for example by heightening political discontent and/or affecting an incumbent's vote-share in "hybrid" regimes. Section 5.7 applies these findings to account for the political dynamics in Egypt and Jordan during the Arab Spring. Section 5.8 concludes by challenging the widespread embrace of remittances as a panacea for the problems plaguing developing countries. In nondemocracies, the economic benefits should be weighed against the political costs.

5.1 REMITTANCES AND POLITICS

To date, scholars have identified two broad channels through which remittances might affect politics: political engagement and public finance. This section briefly discusses each.[3]

---

[3] For a broader discussion of migration and politics, see Meseguer and Burgess (2014) and Mosley and Singer (2015).

Remittances and Political Engagement

Modernization theories stipulate that rising individual incomes and economic growth should heighten demands for democracy. As a source of household income, remittances therefore have the capacity to transform political preferences and participation.[4] In democracies, the evidence linking remittances to political participation is mixed. Survey data from Mexico, for instance, suggest that remittance recipients are less likely to vote and seek out political information, but are more likely to engage in local nonelectoral organizations.[5] This voter apathy may, however, reflect support for the incumbent's policies among remittance-receiving households.[6]

Even politically engaged remittance recipients may choose to support an incumbent government. This voting calculus can be affected by the prevalence of clientelism in electoral campaigns (e.g., vote-buying), especially in many developing countries. With such a setting in mind, Ahmed (2017a) develops a model highlighting two countervailing effects associated with remittances and electoral politics. On the one hand, remittance recipients may attribute their additional income favorably to the performance of the incumbent government (whether this is deserved or not). Yet, on the other hand, the addition of remittance income can raise the marginal costs for political parties – including the incumbent – to "buy" the electoral support of remittance recipients. Ahmed reconciles these divergent effects and shows – both theoretically and empirically with survey data from Latin America – that remittance recipients are more likely to vote for an incumbent they dislike relative to nonrecipients.

While survey evidence evaluating the effects of remittances on political participation in *autocracies* remains scant, remittances may also augment the relationship between the government and voters in electoral autocracies. Escriba-Folch et al. (2015) posit that remittance recipients are less dependent on patronage networks in electoral autocracies, especially in party-based dictatorships, and are less likely to support the incumbent. In a cross-national setting, these scholars show that remittances received in party-based dictatorships increase the likelihood of democratization via an electoral mechanism. These findings suggest that remittances may potentially undermine autocracy by decreasing an incumbent's vote-share (at least in party-based dictatorships).

Remittances and Public Finance

While remittances can potentially affect the political behavior of individual recipients, their large volume at the national level has the potential to affect

---

[4] Bearce and Park (2017) summarize this argument.
[5] Goodman and Hiskey (2008); Perez-Armendariz and Crow (2010). [6] Ahmed (2017a).

a government's public finances. Migrant remittance income, however, does not directly accrue to a government, especially because such income is poorly tracked and is essentially untaxed by governments in developing countries.[7] Consequently, remittance income rarely directly enters a government's revenue base. Nevertheless, remittances may do so indirectly via a substitution effect.[8]

As argued in Chapter 3, the substitution effect posits that a government adjusts its composition of public finances in response to untaxed remittance income. A household can spend some fraction of its remittance income on a substitutable welfare good (i.e., a good that can be provided by the government or household, such as basic health care). In response, a government can reduce its own provision of that welfare good and divert its "free" resources to patronage and repression. Moreover, the incentive for a leader to engage in this strategy is larger in nondemocracies (as argued in Chapter 3). Thus, remittance income can sustain dictatorial rule.

In addition to shifting the composition of government spending, higher levels of aggregate remittances could expand *total* government spending by reducing the cost of sovereign borrowing. This potentially complementary mechanism relies on the notion that remittances help ease sovereign borrowing costs by enhancing a country's creditworthiness.[9] For example, in a cross-national study Singer (2012) finds that remittances can lower a country's default risk (by lowering its sovereign debt spread) and consequently increase government consumption. Thus, by facilitating a country's access to international credit, remittances may allow a government to expand its total spending, including expenditures on policies that benefit the government.

Given the various channels through which remittances might affect autocracy, I now turn to the data to discern the veracity of such an empirical relationship.

5.2 QUASI-NATURAL EXPERIMENT

Endogeneity Problem

Endogeneity bias arising primarily from measurement error and reverse causality plagues empirical evaluations of remittances on politics. Regarding measurement error, officially recorded inflows of remittances tend to underreport actual flows, as poorer countries lack the bureaucratic and technical capacity to properly track remittance inflows.[10] Thus, measurement error will tend to attenuate the estimated effects of remittances toward zero. Regarding reverse causality, the decision to emigrate and remit earnings is often driven by a lack of economic opportunities in the home country, many of which

---

[7] Chami et al. (2008). [8] Abdih et al. (2012); Ahmed (2012). [9] Ratha (2007).
[10] de la Martinez (2005).

5.2 Quasi-natural Experiment

are nondemocratic.[11] This correlation between remittances and autocracy will upward bias the effect of remittances on authoritarian politics.

Identification Strategy

One strategy to mitigate this endogeneity problem is to identify an instrumental variable for remittances. I exploit plausibly exogenous variation in the world price of oil *interacted with* a Muslim country's distance to Mecca to construct a powerful cross-national and time-varying instrumental variable (IV) for remittances sent from oil-producing countries in the Persian Gulf to poor, *non–oil*-producing Muslim countries.[12]

The IV exploits two sources of variation. After the 1973 oil crisis, labor from different countries in North Africa, South Asia, and Middle East migrated in great numbers to the oil-exporting countries in the Persian Gulf. The first wave of workers (totaling about 500,000) migrated from non–oil-producing Gulf countries, such as Jordan, Palestine, and Yemen. In the latter part of the decade, Gulf countries began to recruit a large number of South Asian workers from India, Pakistan, and Bangladesh. For example, it is estimated that the number of Pakistani workers jumped from roughly 500,000 in 1975 to more than 1.25 million in 1979. By the early 1980s, there may have been some 3.5 million to 4.65 million migrants, in a combined labor force of 9 million to 10.2 million workers.[13] This large movement of labor generated large capital flows in the form of worker remittances from Gulf oil producers to a variety of non–oil-producing labor exporting countries, such as Jordan, Mali, and Pakistan.

Two stylized facts underlie the instrumental variable. First, the amount of aggregate remittances received by poor, non–oil–producing Muslim countries tracks the world price of oil. As Figure 5.1 shows, as the price of oil began to rise in 1974, remittance inflows to poor non–oil-producing Muslim countries rose sharply.[14] This level of remittance remained high through the early 1980s and

---

[11] Dictators may actively encourage emigration. For example, under the dictatorial rule of Marcos, the Philippines actively sought to "export" labor to the Persian Gulf and the United States to promote remittances as a survival strategy. The instrumental variable (described in the next subsection) addresses this case of self-selection by leveraging variation in world oil prices (which is exogenous to the labor-exporting policies of Marcos and those of all other remittance-receiving countries) to explain the variation in a country's aggregate remittances.

[12] This identification strategy follows Ahmed (2013), which examines the effect of remittances on governance (corruption, public spending) over the period 1984–2004. The analysis in this chapter begins in 1972.

[13] Choucri (1986).

[14] The y-axis on the left scale measures the total remittances as a share of total GDP for a set of non–oil-producing Muslim countries. A country is defined as Muslim if at least 70 percent of the population identify with the Islamic faith. Moreover, remittances received in Muslim countries were more responsive to oil prices. The correlation between oil price and average remittances in Muslim countries is 0.54, compared to only 0.26 in non-Muslim countries.

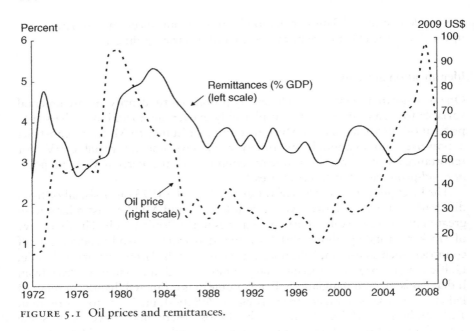
FIGURE 5.1 Oil prices and remittances.

then began to fall as the price of oil plummeted. Since the 1990s, remittance flows have tended to be less volatile but still tend to co-move with the price of oil. As supply decisions in oil producers and demand conditions in large (industrialized and rich) economies determine oil prices, *the price of oil provides a plausibly exogenous source of variation in remittance flows* that is unrelated to the economic, political, and social conditions in remittance-receiving countries.

The second stylized fact is that remittances inflows to non–oil-producing Muslim countries are inversely related to each country's distance from the Persian Gulf. Countries closer to oil-producing Gulf economies experience greater outward migration and subsequently higher remittance inflows. To account for this fixed cost of migration to the oil-producing economies in the Persian Gulf, I use a country's distance to Mecca. Table 5.1 shows that non–oil-producing Muslim countries closer to Mecca tended to receive higher remittances (as a share of gross domestic product [GDP]) and these countries tend to be autocratic. For instance, over the sample period Jordan (765 miles from Mecca) on average received remittances equal to 18.4 percent of GDP. In contrast, Bangladesh (3,212 miles from Mecca) on average received remittances equal to 4 percent of GDP.

Instrumental Variable. Given these two stylized facts, I interact the price of oil with a Muslim country's distance from Mecca as an instrument variable (IV) for remittances. The reduced form two-stage regression setup is

5.2 Quasi-natural Experiment

TABLE 5.1 *Remittances and distance to Mecca*

Distance to Mecca	Countries	Remittances (% GDP)	POLITY2
Less than 1,000 miles	Eritrea, Sudan, Djibouti Jordan, Lebanon	9.31	−3.92
1,000 to 2,000 miles	Turkey, Somalia, Morocco Pakistan, Niger, Mali	4.33	−1.45
More than 2,000 miles	Bangladesh Guinea, Senegal	3.08	−0.96

Note: Sample of poor, non–oil-producing Muslim countries.

$$\text{Firststage}: REMIT_{it} = \alpha + \beta \left(DIST_i^* \, p(oil)_t \right) + \gamma X_{it} + \delta_t + \delta_i + \varepsilon_{it}$$

$$\text{Secondstage}: POLITICS_{it} = a + b^* REMIT_{it} + c^* X_{it} + d^* Y_t + f^* D_i + u_{it}$$

In the second stage, $POLITICS_{it}$ measures various political outcomes – executive constraints, POLITY score, leader exit – in country i in year t. $REMIT_{it}$ is each country's officially recorded inflows of remittances. $DIST_i$ measures the log distance of a poor, non–oil-producing Muslim country to Mecca. A country is classified as Muslim if at least 70 percent of its population identifies with the Islamic faith.[15] As discussed shortly, this threshold ensures that the treatment and control groups of countries are "comparable" prior to the application of the treatment (i.e., oil price–induced remittances beginning around 1972). X_{it} is a set of time-varying covariates, Y_t is a year trend, and D_i are dummies for each country.

The instrument is exogenous because the regressions control for the main effect of the potentially endogenous variable (i.e., distance to Mecca) with the inclusion of country fixed effects, D_i.[16] The estimating sample is restricted to poor (as defined by the World Bank) and non–oil-producing countries (as classified by British Petroleum). As oil prices are likely to have a direct effect on economic and political conditions in oil producers these countries (e.g., Mexico) are excluded from the analysis, so as not to bias the findings

---

[15] This percentage is based on estimates from the CIA world factbook, available at www.cia.gov/library/publications/the-world-factbook/
[16] The identifying assumption is that the endogenous variable and the outcome of interest are jointly independent of the "exogenous" variable. For a more technical discussion, see section 2.3.4 of Angrist and Kruegler (1999). The inclusion of country fixed effects (D_i) accounts for the potentially "endogenous" components of the instrument because distance to Mecca is country specific and time invariant.

associated with remittances.[17] I estimate both stages using ordinary least squares (OLS) and to take into account serial correlation, I cluster the standard errors at the country level.

Exclusion Restriction. The validity of this chapter's identification strategy relies on whether the exclusion restriction is satisfied: oil prices and their interaction with a country's distance to Mecca affect *POLITICS* via remittances *only*. While the inclusion of country fixed effects accounts for *any* time-invariant effect on *POLITICS* (e.g., geographic proximity to the Middle East, Islamic identity, culture, colonial legacy), time-varying factors correlated with oil prices might contaminate the identification strategy. In particular, during the period of high oil prices Arab oil producers disbursed large amounts of foreign aid to poor, non–oil-producing Muslim countries, and several countries (e.g., Saudi Arabia, Iran, and Libya) sought to use their oil wealth to fund insurgencies in other countries.[18] To account for these effects, all the specifications control for foreign aid and the incidence of civil war. Finally, including GDP per capita growth controls for the effects of oil prices on economic growth (and their subsequent effect on political stability via discontent, for example).

Causal Inference. In the second-stage regression, the coefficient on remittances will measure the average treatment effect for a group of poor, non–oil-producing Muslim countries that are nondemocratic. Over the sample period, the average POLITY score for these countries is –1.2, falling well shy of democratic governance (POLITY >+6). Within the treatment group, these countries vary in the quality of governance. For instance, Jordan is a monarchy with an average POLITY score of –5.2. In contrast, Bangladesh and Turkey fluctuated between episodes of autocracy and weak democracy. Bangladesh swings from a POLITY score of –7 from 1975 to 1985 to +6 for most years since 1991, while Turkey moves between a low POLITY score of –5 to a high of +9. This rich variation in the types of autocracies implies the instrumental variable will gauge a more precise average treatment effect of remittances received in poor, non–oil-producing Muslim autocracies.

Data

Measuring Authoritarian Politics. Throughout history, autocrats and democratic leaders differ in the constraints they face.[19] Thus, I use the executive constraints index (*XCONST*) from the POLITY IV data set as the primary measure of *POLITICS*.[20] *XCONST* measures the extent of institutionalized constraints on the decision-making powers of the chief

---

[17] See Ramsay (2011) on how oil prices affect democracy in oil producers.
[18] Lemarchand (1988); Kepel (2002); Neumayer (2003).   [19] Gledistich and Ward (1997).
[20] Marshall et al. (2015).

5.2 Quasi-natural Experiment

TABLE 5.2 *Executive constraints and political survival*

XCONST	Degree of executive authority	POLITY2	Years in office	EXIT
	(1)	(2)	(3)	(4)
1	Unlimited	−7.34	11.6	0.07
2	Intermediate	−5.67	11.3	0.08
3	Slight to moderate limitations	−3.41	10.97	0.08
4	Intermediate	3.73	10.58	0.06
5	Substantial limitation	5.98	5.41	0.13
6	Intermediate	7.5	4.06	0.15
7	Executive parity	8.26	3.81	0.19

Notes: Each column reports the average value for that variable at the corresponding level (index value) of executive constraints. For example, in column 4 *EXIT* is the average probability a government (leader) will lose office in a given year corresponding to that level of executive constraints.

executives, whether they are individuals or collectives. The index has been used widely to capture the "predatory" nature of a state and cross-national differences in the quality of governance.[21]

The executive index ranges on a 7-point scale (1 to 7), where lower values imply a less constrained executive. A value of 1 corresponds to a leader with "unlimited authority"; a score of 7 implies the leader is "subordinate" to the governed population. In the estimating sample, 13 percent of country–year observations have leaders with unlimited authority, 32 percent with "slight to moderate constraints" ($XCONST = 2, 3, 4$), 20 percent with "substantial" limitations ($XCONST = 5$), and the remaining 35 percent with high constraints ($XCONST = 6, 7$). These constraints are correlated with a country's regime type and affect a government's longevity. For the estimating sample of countries, Table 5.2 shows that leaders who face lower constraints govern in less democratic countries (i.e., lower POLITY scores), last longer in power, and are less likely to lose office.

Independent Variables. I measure the key independent variable, workers' remittances, using the World Development Indicators (WDIs). It measures officially recorded flows of remittances and will tend to understate actual remittances because a large share of these capital flows go through back channels and/or are difficult to monitor by poor governments.[22] I measure the annual value of aggregate remittances (in log units) received by a country.[23] In the estimating sample, the typical country receives remittances equal to 18.4 log

---

[21] Gledistich and Ward (1997); Bueno de Mesquita et al. (2003).   [22] de Luna Martinez (2005).
[23] The results are robust, with remittances measured as a share of GDP.

units (equivalent to about 4.2 of its GDP), although there is wide cross-sectional and temporal variation. For instance, the bottom decile of observations records aggregate remittances that are less than 0.1 percent of GDP. In contrast, the top decile of observations records remittances that exceed 11 percent of GDP. In the top decile, aggregate remittances exceed $2 billion (2000 US$) annually.

Variables measuring economic, demographic, and country-specific factors are also likely to affect governance. The economic (real GDP per capita, real GDP per capita growth) and demographic (log population) variables are from the WDI. The typical country is poor, with real GDP per capita equal to around $1,350 with modest economic growth (= 1.8 percent per annum). The main results control for observed and unobserved time-invariant country characteristics (using country fixed effects). The inclusion of country fixed effects accounts for observable country characteristics such as a country's legal origin, colonial heritage, religious tradition, and ethnic-linguistic fractionalization. The inclusion of country fixed effects implies that the estimated effects will gauge within country variation in *POLITICS*.

5.3 REMITTANCES ENHANCE AUTOCRATIC POWER

Establishing Validity

Pretreatment Balance. As the previous section makes clear, the instrument identifies the effect of oil price–induced remittances in the treatment group of poor, non–oil-producing Muslim countries only. Thus poor, non–oil-producing *non-Muslim* countries constitute the control (or reference) group. A country qualifies as Muslim if 70 percent of its population identify with Islam. This threshold ensures that the treatment and control groups of countries are "comparable" *prior* to the application of the treatment (i.e., oil price–induced remittances beginning around 1972) and strengthens the validity of the causal inferences.

Table 5.3 shows that the treatment and control groups are "balanced" on a number of observable country characteristics prior to the treatment period. This is evident because the difference in group means between Muslim and non-Muslim remittance recipients across each observable characteristic (column 3) is not statistically significantly different from zero (i.e., the *p*-values are not statistically significant in column 5). This means that Muslim and non-Muslim non–oil producers do *not* differ from each other on important economic and demographic conditions prior to the first oil shock. This implies that subsequent divergences in governance between these groups of countries may be attributable to remittances, while *controlling* for potential confounding effects, such as income, economic growth, foreign aid, etc.

The Determinants of Remittances. Specification 1 in Table 5.4 reports the first stage regression. These fixed effect specifications control for economic and

5.3 Remittances Enhance Autocratic Power

TABLE 5.3 *Pretreatment characteristics between Muslim and non-Muslim countries prior to oil price shock*

	Non-Muslim	Muslim	Difference	Std. error	p-value
	(1)	(2)	(3)	(4)	(5)
Executive constraints	3.17	2.27	0.9	0.62	0.16
Log GDP per capita (2000 US$)	6.66	6.03	0.63	0.33	0.11
Log population	15.29	15.66	−0.37	0.36	0.31
Foreign aid (% GDP)	2.12	3.12	−1	0.89	0.27
Civil war	0.14	0.2	−0.06	0.11	0.56

Notes: Mean values for poor, non–oil-producing Muslim and non-Muslim countries. In column (5), the *p*-value is based on a *t*-test of differences in mean. Column 4 reports the standard error for the difference in group means (reported in column 3).

demographic characteristics and explain 85 percent of the variation in remittance inflows. In column (1), the coefficient on the instrument is 0.008 and is significant at the 1% level. For the typical poor, Muslim non–oil producer (which is about 2,000 miles from Mecca), this coefficient implies that a $10 increase in the world price of oil raises inflows of log remittances by 0.61 log unit (this is equivalent to 0.5 percent of GDP). Between 1976 and 2009, oil prices ranged from $17 (in 1998) to $97 (in 2008) per barrel. Such a movement in oil prices represents a shock to remittances equal to 2.11 log units (or about 4% of GDP).

The control variables have the expected signs (not reported). Richer and more populated tend to receive fewer remittances. Growth exhibits a positive and statistically effect on remittances, which is surprising because remittances tend to be countercyclical. The *F*-statistic on the instrument (= 15.3) far exceeds the conservative threshold of "weak" instruments of 9.6 suggested by Stock et al. (2002). This implies the second stage effects on remittances on *POLITICS* can be interpreted as *causal*.

Remittances Increase Autocratic Power

Armed with this strong instrumental variable, columns 2 and 3 in Table 5.4 report the effect of instrumented remittances on *XCONST* and *POLITY*. In column 2, the coefficient estimate implies that a one log unit increase in remittances lowers executive constraints by nearly two index points. This effect is substantively meaningful. For instance, a 1 standard deviation in log remittances corresponds to a 2.1 standard deviation decrease in *XCONST*. In the estimating sample this is equivalent to moving from a country with relatively high executive constraints such as Botswana (*XCONST* = 6) to a country such

TABLE 5.4 *Remittances strengthen authoritarian politics*

	First stage	Second stage		
	Log remittances	Executive constraints	POLITY	Turnover
Method of estimation:	OLS	2SLS	2SLS	IV Probit
	(1)	(2)	(3)	(4)
Log distance to Mecca × Oil price ($t-1$)	0.008 (0.002)***			
Log remittances (2000 US$)		−1.785 (0.845)**	−4.037 (2.321)*	−0.249 (0.090)***
No. of observations	1,392	1,392	1,434	1,425
R^2/log likelihood	0.85	0.2	0.31	−3,393.34
F-statistic on instrument	15.27	15.27	15.52	16.79

Notes: Robust standard errors clustered by country are reported in parentheses. *, **, *** = significant at 10%, 5%, and 1% respectively. All specifications include country fixed effects, a year trend, and the following time-varying country characteristics: log GDP per capita (2000 US $), real GDP per capita growth (% annual), log population, foreign aid (% GDP), and the incidence of civil war. These coefficients and a constant are not reported. Column 4 also controls for the duration of the incumbent government (leader). OLS, ordinary least squares; 2SLS, two-stage least squares.

as Niger or Mali (both with $XCONST = 3$). As these estimates gauge the average treatment effect for a group of autocracies, the results imply that *remittances expand the powers of existing dictators.*

This downward movement in executive constraints produces a substantial loss in aggregate economic welfare. For example, Gledistich and Ward (1997) show that a decline in $XCONST$ implies a movement away from democracy and Acemoglu and Johnson (2005) argue that a lower level of $XCONST$ corresponds to less secure property rights and an environment conducive to lower economic growth, investment, and financial development. More broadly, remittances tend to lower a country's overall quality of democracy. The estimated effect in column 3 implies that a modest one log unit increase in aggregate remittances lowers a country's POLITY score by nearly 4 index points.

These core results, which show that remittances cause a reduction in executive constraints, are robust to a variety of specification checks and concerns with unobserved heterogeneity. Country "jackknife" regressions show the findings are not unduly influenced by remittances from any particular country in the treatment group, and the core finding result holds using contemporaneous oil prices (i.e., oil price in year t) to instrument for remittances. The results are also robust to unobserved temporal effects and differential trends.[24] For example, remittances reduce executive constraints in specifications that control for the Cold War period.[25]

Unobserved differential trends between the control and treatment groups of countries may also affect executive constraints, independently of remittance inflows. For example, demographic changes since the 1970s across Muslim and non-Muslim countries have not been identical (e.g., due to higher fertility rates and lower income growth). Moreover, religion has become an increasingly prominent feature of politics in Muslim countries (relative to non-Muslim countries). These unobserved demographic and "Islamization" dynamics may affect the decision-making calculus of leaders. To address these worries, I reestimate the baseline specification controlling for a *Muslim* × *Year* interaction. In this specification, the substantive effect of remittances on *XCONST* remains unchanged.

5.4 POLITICAL SURVIVAL

From the perspective of a leader, a reduction in political constraints (induced by higher remittances) is a rational strategy if it extends his time in power. This seems to be the case. Column 4 in Table 5.4 shows the causal effect of remittances on government termination (*EXIT*), controlling for economic and demographic characteristics, *and* the number of years that government has lasted in power (*Years in office*).[26] The dependent variable, *EXIT*, is a binary variable equal to 1 if the ruling government in country i loses office in year t, and 0 if that incumbent remains in power.[27]

Because the outcome variable is binary, the regression is estimated via instrumental probit, where the first-stage regression is estimated via OLS and the second stage via probit. In the second stage, a negative coefficient on a covariate implies that variable lowers the probability of government turnover. Based on this interpretation, the marginal effect associated with the coefficient estimate in column 4 implies that a modest 1 log unit increase in remittances lowers the likelihood of turnover by 25 percent. This is equivalent to about a 200 percent reduction in the baseline probability that a government will lose power.

---

[24] These results are available with the book's replication materials.
[25] In this specification, I control for a Cold War dummy that is equal to 1 in the years prior to 1990, and 0 after 1990.
[26] From an econometric standpoint, *Years in office* controls for duration dependence. From a political economy perspective, this variable accounts for a government's incumbency advantage.
[27] *Years in office* and *EXIT* are drawn from the Database of Political Institutions (Beck et al. 2010).

5.5 EVALUATING THE CHANNEL

Evidence of a Substitution Effect

Chapter 3 theorizes that remittances can fund authoritarian politics via a substitution effect whereby a government can divert some expenditures from welfare goods to patronage. Table 5.5 provides evidence of this effect. To measure government patronage, I use the *share of government expenditures* devoted to government employee compensation. Government patronage, especially in many nondemocracies, tends to be highly correlated with public sector employment, as it frequently reflects the government's incentives to channel spending to targeted constituencies.[28] Moreover, in autocracies, a large portion of these workers are likely to be within the government's "inner circle."[29] Thus, higher public sector compensation (total employment and/or higher wages) provides an objective and observable measure of government patronage across countries. I use a government's expenditure on subsidies and transfers as the broadest measure of welfare spending. Consistent, reliable, and comparable cross-national data on public finance is available only from 1990 onwards (thus the smaller estimating samples in Table 5.5).

The results in Table 5.5 demonstrate a substitution effect: remittances permit governments to increase their provision of patronage by reducing welfare spending. In column 1, a 1 log unit increase in remittances permits a government to allocate an additional 11.4 percent of government expenditures to public sector compensation. The increase in patronage in column 1 is due in large part to a reduction in the provision of welfare goods, as a 1 log unit increase in aggregate remittances lowers spending on subsidies and transfers by 8.7 percentage points of government expenditures (column 2).

Columns 3 and 4 evaluate two additional features associated with the substitution effect. Column 3 shows that remittances are negatively correlated with taxes collected from household income, profits, and capital gains (as a share of total taxes). This supports the conjecture that remittances do *not directly* add to a government's tax base. Rather, remittances seem to decrease the government's tax effort. Another feature associated with the substitution effect is the notion that remittances increase household consumption, which in turn permits a government to reduce its expenditures on welfare goods. Column 4 shows that remittances increase total household consumption, which includes expenditures on welfare goods, such as health care and education. Finally, to validate the underlying notion that a larger public sector is associated with authoritarian politics, column 5 regresses *XCONST* on public sector employment. The negative coefficient on *XCONST* implies that a government with fewer executive constraints tends to allocate a larger share of its budget on patronage.

---

[28] Keefer (2007).   [29] Bueno de Mesquita et al. (2003).

5.5 Evaluating the Channel

TABLE 5.5 *Remittances and public finances*

	Patronage	Welfare goods	Tax effort	Household consumption	Patronage
Method of estimation:	2SLS	2SLS	2SLS	2SLS	OLS
	(1)	(2)	(3)	(4)	(5)
Log remittances (2000 US$)	11.438 (6.816)*	−8.704 (4.865)*	−4.97 (2.902)*	6.662 (2.575)***	
Executive constraints					−1.943 (1.653)
No. of observations	474	472	516	1402	493

Notes: Robust standard errors clustered by country are reported in parentheses. *, **, *** = significant at 10%, 5%, and 1% respectively. Columns 1–4 are estimated via two-stage least squares. Column 5 is estimated via ordinary least squares (OLS). All regressions include the following additional time-varying covariates: log GDP per capita (2000 US$), GDP per capita growth (% annual), log population, foreign aid (% GDP), civil war, a year trend, and country fixed effects or a Muslim indicator in the smaller samples. These coefficients and a constant are not reported. In columns 1 and 5, patronage measures compensation on public sector employment (% government expenditures). In column 2, welfare goods measures expenditures on subsidies and transfers (% government expenditures). In column 3, tax effort measures taxes collected from household income, profits, and capital gains (% of total taxes). In column 4, household consumption measures final household expenditures (% GDP).

Remittances and Corruption

Politically, remittances can fund corruption. And as described in Chapter 3, engaging in corruption can be an effective strategy of political survival for an autocrat.[30] Corruption is typically defined as the "misuse of public office for private gain," where the private gain may accrue to either the individual public official or to the groups of parties to which he belongs, such as his political party or governing coalition.[31] This definition is quite broad and can entail various forms of corruption, such as payments from private citizens to public officials (bribery) and transfers from the government to targeted groups (patronage).

---

[30] As noted in Chapter 3, according to the selectorate model of political survival (Bueno de Mesquita et al. 2003), governments in autocracies are likely to foster corruption for at least three reasons. First, eliminating corruption and encouraging institutions that promote the rule of law are public goods. Leaders in autocracies have few incentives to find and eliminate corruption. Second, leaders can provide benefits by granting the supporters the right to expropriate resources from themselves. Thus, autocratic leaders might encourage corrupt practices as a reward mechanism. Third, the prevalence of kleptocracy in autocracies frequently allows leaders to siphon off resources for pet projects. "Ruling to steal" constitutes a form of corruption.

[31] Bardhan (1997).

The conjecture that remittances ease the pressure of government to provide welfare goods in order to fund patronage requires a measure of *patronage-based* corruption. One such measure is the International Country Risk Guide (ICRG) corruption index. Unlike other measures that focus primarily on bribery, the ICRG corruption index explicitly strives to quantify government patronage. The ICRG is clear that its corruption measure "is more concerned with actual or potential corruption in the form of excessive patronage, nepotism, job reservations, 'favor-for-favors', secret party funding, and suspiciously close ties between ties between politics and business." To the extent that government patronage is often quite difficult to reliably quantify for a large group of countries over time, the ICRG corruption measure captures the trade-off between patronage and the provision of welfare goods (see Table 5.6, column 2). For instance, the corruption index is negatively correlated with expenditures on public health care, education, and social contributions and positively correlated with the size of the public sector.[32] The "raw" index ranges from a score of 0 (high corruption) to 6 (low corruption) and is available on a monthly basis from 1984 onwards. I take annual averages and *rescale* the index so that higher values correspond to greater corruption, CORRUPTION (i.e., 6 = Highest corruption, 0 = Lowest corruption).

Table 5.6 examines the determinants of CORRUPTION. I first verify that CORRUPTION is in fact correlated with nondemocratic governance. In column 1, the negative coefficient on POLITY implies that CORRUPTION is lower in countries with higher quality democratic institutions.[33] And on examining the composition of public finances, governments that exert greater effort on the provision of welfare goods tend to be less corrupt (column 2). Columns 3 and 4 show that instrumented remittances cause greater corruption. Column 3 reports the impact of remittances in a specification without any country fixed effects but that does correctly control for the time-invariant component of the instrumental variable (log distance to Mecca).[34] The coefficient on remittances is positive and statistically significant and implies that a 1 standard deviation increase in remittances raises corruption by more than 1 full index point (which is equivalent to a 1 standard deviation in CORRUPTION).[35] Controlling for country fixed effects – which subsumes the effect of distance to Mecca – does not attenuate this effect (column 4). In fact, the coefficient estimate is slightly larger in magnitude (coefficient = 0.15). In combination, the results in Table 5.6 show that patronage-based corruption is correlated with autocracy and remittances cause corruption.

---

[32] Ahmed (2013) shows this. The results in Table 5.6 are drawn from Ahmed (2013).
[33] Corruption is also negatively correlated with XCONST and the likelihood an autocratic government will lose power.
[34] The coefficient on log distance to Mecca is negative and significant, which is consistent with the finding that countries closer to the "center" of Islam have historically exhibited nondemocratic governance (Chaney 2012).
[35] For the sample in this specification, the standard deviation for log remittances is 10.8.

5.5 Evaluating the Channel

TABLE 5.6 *Remittances raise corruption*

Dependent variable:	Corruption (0 = Lowest, 6 = Highest)			
Method of estimation:	OLS (1)	OLS (2)	2SLS (3)	2SLS (4)
POLITY	−0.03 (0.010)***			
Welfare goods		−0.007 (0.004)*		
Log remittances (2000 US$)			0.117 (0.042)***	0.157 (0.076)**
Log distance to Mecca			−0.597 (0.204)***	
Country fixed effects				Yes
No. of observations	1,678	528	878	878

Notes: Robust standard errors clustered by country are reported in parentheses. *, **, *** = significant at 10%, 5%, and 1% respectively. Columns 1 and 2 are estimated via ordinary least squares. Columns 3 and 4 are estimated via two-stage least squares. All regressions include the following additional time-varying covariates: log GDP per capita (2000 US$), GDP per capita growth (% annual), log population, foreign aid (% GDP), civil war, a year trend, and country fixed effects. These coefficients and a constant are not reported. In column 2, welfare goods are government expenditures on subsidies and transfers (% total government expenditures).

Discounting Other Channels

Taken together, Tables 5.4, 5.5, and 5.6 provide compelling evidence that remittances foster autocracy by decreasing a government's provision of welfare goods in favor of patronage. In this subsection, I evaluate whether the other channels discussed in Section 5.1 (e.g., sovereign borrowing, "modernization" effects, and political discontent) are also viable channels through which remittances might affect autocracy. The estimates in Table 5.7 show that there is much weaker evidence of these other channels.

Starting with the sovereign borrowing mechanism, columns 1–3 in Table 5.7 show that remittances do not affect a country's access to foreign assets or total government consumption. To capture a country's access to foreign capital, I use a country's net capital balance (% GDP) as the broadest measure of a country's acquisition of foreign assets. In column 1 remittances have a small positive but statistically insignificant effect on a country's net capital account balance.[36] This positive coefficient (= 0.59) implies that remittances do somewhat increase

---

[36] The capital account is one of two primary components of the balance of payments. Whereas the current account reflects a nation's net income, the capital account reflects a net change in a nation's ownership of assets. Remittance income accrues to the current account.

TABLE 5.7 *Discounting other channels*

Channel:	Sovereign borrowing			"Modernization" effects		Political discontent		Elections
Dependent variable:	Capital account (% GDP)	Imports (% GDP)	Government consumption (% GDP)	Log GDP per capita	GDPPC growth	Riots	Antigovernment demonstrations	Vote share
	(1)	(2)	(3)	(4)	(5)	(6)	(7)	(8)
Log remittances	0.592	10.448	−0.192	0.102	−1.557	0.221	0.08	19.621
(2000 US$)	(0.530)	(3.047)***	(0.980)	(0.069)	(1.588)	(0.459)	(0.439)	(32.797)
No. of observations	1,343	1,367	1,405	1,451	1,449	1,433	1,434	441

Notes: Estimation via two-stage least squares (2SLS). Robust standard errors clustered by country are reported in parentheses. *, **, *** = significant at 10%, 5%, and 1% respectively. All regressions control for log GDP per capita (2000 US$), GDP per capita growth (% annual), log population, foreign aid (% GDP), incidence of civil conflict, total trade (% GDP), inflation, year trend, and country fixed effects. These coefficients and a constant are not reported. The dependent variables (by column) are net capital account as a share of GDP (column 1), final government consumption expenditures as a share of GDP (column 2), the number of riots (column 6), number of antigovernment demonstrations (column 7), and incumbent vote share from recent presidential and/or legislative election (column 8). In column 8, the sample is restricted to years of a presidential and/or legislative election only.

5.5 Evaluating the Channel

the flow of foreign assets into a country. However, because remittance income is accounted for on the other side of the balance of payments (i.e., the current account), *if* remittance income did lower the cost of acquiring foreign assets, then the coefficient estimate should exceed 1.[37] Rather the coefficient estimate is less than 1, implying that some remittance income is "leaking out" of the economy.

The positive and statistically significant coefficient in column 2 shows that imports are a likely source of this leakage, as remittances raise a country's imports of goods and services (which require an outflow of a country's assets in order to buy those foreign goods and services). To the extent that access to foreign assets affects the *total size* of government spending, remittances seem to *reduce* a government's total consumption (column 3). While the estimate is statistically insignificant, the negative coefficient (=−0.19) is not consistent with Singer's (2012) central hypothesis. On balance, these weak effects suggest that remittances neither expand a government's access of foreign assets nor increase the government's total expenditures.

There is also inconclusive evidence in support of modernization theory (columns 4 to 7). This theory stipulates that rising individual incomes and economic growth (both attributable to household remittance income) should encourage political liberalization and trigger demands for political change. While remittances are positively correlated with higher per capita income (column 4), the effect is statistically indistinguishable from zero. Moreover, remittances exhibit a negative but statistically insignificant effect on economic growth (column 5). Given these findings, remittances are unlikely to engender strong demands for political change. Columns 6 and 7 show this to be the case. While remittances are associated with a slight bump in the number of riots (column 6) and antigovernment demonstrations (column 7), the effects are not statistically significant.[38]

Finally, in electoral settings, remittances are positively correlated with the total vote share of government parties (column 8).[39] This increase in electoral support may be due to the redirecting of funds toward patronage (core supporters) and is at odds with the clientelistic mechanism advanced by Escriba-Folch et al. (2015). This effect, however, is not statistically significant and should be interpreted as suggestive, at best.

---

[37] Because the balance of payments is an accounting identity, a country's net current account balance must equal a country's net capital balance.

[38] A riot refers to any violent demonstration or clash of more than 100 citizens involving the use of physical force. An antigovernment demonstration refers to any strike of at least 1,000 individuals that is aimed at national government policies or authority. These variables are drawn from Banks (2010).

[39] Specification 8 is restricted to observations from legislative and/or presidential election years, hence the small sample size. The vote-share data are from Beck et al. (2010).

5.6 REMITTANCES AND THE ARAB SPRING

The notion that remittance income can constitute a type of government rent that strengthens authoritarian rule may be relevant for understanding recent developments in the Arab world. Trends in remittance income provide an additional, and to the best of my knowledge, unexplored, source of instability underlying the 2011 Arab Spring. During this three-month period, dictators in several Muslim countries lost power (e.g., Egypt, Tunisia, Yemen), while some others survived (e.g., Bahrain, Jordan).[40] In some of these countries, aggregate remittances comprised a nontrivial share of national income.[41] In particular, the divergent trajectory of remittance inflows in Egypt and Jordan provide evidence of a correlation between remittances and authoritarian survival during the Arab Spring.

Patronage Politics. In Egypt and Jordan patronage politics underlies authoritarian rule. In both countries, the incumbent relies on external rents to survive. Since its inception after World War I, the Hashemite monarchy in Jordan has distributed rents to a small group of coalition members, consisting primarily of Transjordian tribes and business elites.[42] While foreign aid has been a staple source of external assistance to the monarchy since the collapse of the Ottoman Empire, increasingly "indirect" external rents in the form of remittance inflows have enabled the government to finance patronage. As Peters and Moore (2009) note, "authoritarian regimes adapt as different sources of external rent decrease or increase, seeking out new sources of external rent and devising new ways to deliver it to coalition members" (258). Through periods of abrupt demographic changes and intense political violence (both domestically and regionally), Peters and Moore further argue that Jordan's "monarchy, in concert with geopolitically motivated donors, has met these demands by modifying old distributional mechanisms and institutionalizing new venues to take advantage of the international system's provision of economic rents" (257).

Similarly, external rents in Egypt enabled its dictatorship to buy support from the military and to build the apparatus (e.g., secret police) to engage in repressive tactics.[43] An "enduring pattern" of patronage-based politics sustained President Mubarak's governing strategy.[44] Since the late 1970s, foreign aid has played a prominent role in financing Egypt's dictatorship. As Arafat (2009) declares, "for over two decades, Egypt received the second-largest package of foreign aid from the United States (behind Israel), aid that

---

[40] The dictatorial of the Qaddafi regime in Libya also ended in 2011. Unlike the democratic transitions elsewhere at this time, the revolution in Libya required external military assistance from NATO.
[41] In 2009, remittances (as a percentage of GDP) amounted to 15.8 in Jordan, 3.8 in Egypt, 5 in Tunisia, and 5.3 in Yemen.
[42] Brynen (1992); Peters and Moore (2009).   [43] Kassem (1999); Arafat (2009).
[44] Arafat (2009).

5.6 Remittances and the Arab Spring

many Egyptians believe view as supporting the Egypt's authoritarian regime, not its people" (164). Surprisingly, the role of remittance income does not feature prominently at all in existing scholarship in explaining the durability of Egypt's dictatorship, despite such income exceeding foreign aid receipts (see Figure 5.2).

Remittances and Executive Constraints. The relevance of patronage politics and differing trends in external rents provides a backdrop to understand how the Arab Spring unfolded *differently* in Egypt and Jordan. In the lead up to 2011, the political-economic environment in Egypt and Jordan were similar on a number of dimensions. As Figures 5.2 and 5.3 demonstrate, Egypt's President Hosni Mubarak and Jordan's King Abdullah II enjoyed low to moderate constraints on their power ($XCONST = 3$). These figures also describe the trends of foreign aid and remittance inflows in Egypt and Jordan, respectively.

In Jordan, foreign aid receipts have declined substantially since 1980, and with the exception of a spike in aid inflows around the Gulf War I (1990–1991), have stabilized around 5 percent of GDP. While remittance inflows exhibit high annual volatility (as they covary with oil prices) over this period, remittance income has been substantially higher than foreign aid income. Since 2005, remittance inflows have averaged around 19 percent of GDP, which if channeled correctly (as theorized in this book) helped Jordan's monarchy maintain its patronage networks.

Aid and remittance inflows in Egypt exhibit similar dynamics, where remittance income exceeds aid receipts. However, unlike in Jordan, remittance

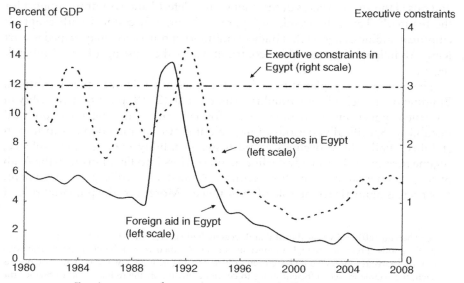

FIGURE 5.2 Foreign rents and executive constraints in Egypt.

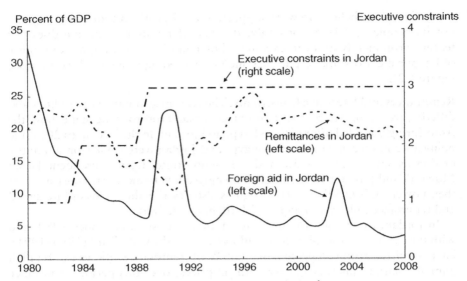

FIGURE 5.3 Foreign rents and executive constraints in Jordan.

income may have been insufficient to offset the downward trajectory of aid inflows into Egypt (from 6 percent of GDP in 1980 to less than 1 percent in 2009). As the momentum of democratic revolution spread from the toppling of the Ben Ali dictatorship in Tunisia to Egypt in February 2011, insufficient external rents in Egypt may have made financing repressive tactics untenable and raised the attractiveness of the military to "defect" and withdraw support for the regime.[45] Thus, the trends in Figures 5.2 and 5.3, coupled with chapter's empirical findings, suggest that higher remittance inflows have may helped thwart King Abdullah's demise, while contributing to the ousting of the Mubarak regime.

This remittance explanation, of course, does not rule out other narratives. Building on models of state fragility, Ahmed et al. (2018) posit that volatility in unearned government income (e.g., foreign aid) can engender political instability. Specifically, periods of high aid inflows can maintain authoritarian stability, while sharp declines in aid foster instability (e.g., greater violence, regime turnover). The steady decline of aid receipts into Egypt prior to the Arab Spring is consistent with this account. Ahmed and colleagues' argument, however, is agnostic to the role of remittances. Moreover, the potential role of

---

[45] See Shehata (2011). Even if the Mubarak regime had successfully contracted its provision of public goods in response to higher remittance inflows (as theorized in this chapter), it is unlikely that those free resources would have been sufficient to augment the low levels of aid receipts. As such, the regime lacked sufficient funds to continue buying off the military and to finance the regime's repressive tactics.

remittances in Middle East politics in 2011 is not mutually exclusive to other potential contributing factors of the Arab Spring, such as those associated with poor economic conditions, demographics, and the media/social networks.[46]

5.7 CONCLUSION

The tremendous rise of remittance flows over the past forty years has been heralded as a "new development mantra" with the means to improve the economic welfare of migrant households in sending countries. These potential improvements in economic well-being may, however, come at the expense of losses in political welfare, by empowering autocrats. In this chapter, I leverage a quasi-natural experimental setting to show that remittances can fund autocracy via a substitution effect. Authoritarian politics is strengthened because remittances permit governments to reduce their expenditures on welfare goods and spend more on patronage that rewards regime supporters (e.g., public sector employment and salaries, corruption).

By extension, therefore, declining remittance income may weaken an autocrat's grip on power, a conjecture that is consistent with the divergent political dynamics of Egypt and Jordan during the 2011 Arab Spring. Buoyant amounts of remittances provided sufficient external rents to sustain patronage in Jordan and hinder the ousting of the ruling monarch during the Arab Spring. This was not the case in Egypt, in which declining remittance income may have weakened the Mubarak regime's ability to fund its various patronage networks, thus making the regime more vulnerable to political upheaval and change.

There are three important takeaways from this chapter's empirical evidence. First, foreign capital does not have to accrue directly to a government for it to constitute a potential source of revenue. Second, the research design evaluates the causal impact of remittances received in a group of autocratic countries. It therefore offers a direct test of the conditional effect of remittances on authoritarian politics: remittances can generate rents in nondemocracies. Third, the presence of "balance" on observable characteristics (e.g., growth, conflict) between Muslim and non-Muslim remittance recipients *prior* to the treatment period (as documented in Table 5.3) suggests that subsequent divergences in governance between these groups of countries *may* be attributed to their differences in remittances, *conditional* on controlling for potential confounding effects, such as income, economic growth, foreign aid, etc. This means that *had* non-Muslim countries also been the beneficiary of oil-price–induced remittance "boom" they *could* have become less democratic as well.

---

[46] See Anderson (2011) and Shehata (2011). For instance, media accounts suggest that poor economic conditions (e.g., high unemployment) fostered domestic discontent, while media/social networks helped motivated and coordinate dissent and protests during the Arab Spring.

6

Foreign Direct Investment in Militarism

For many dictators, their longevity and legitimacy depend on their ability to financially support regime supporters and deliver broad-based economic growth, for when this financial capital and growth disappears (or their prospects), so do most of the reasons for allies to remain loyal.[1] For countries determined to industrialize, such as those in Latin America and East and Southeast Asia, attracting sufficient foreign capital has been instrumental in economic development and legitimizing the state's authoritarian rule.

This chapter presents cross-national evidence that foreign direct investment (FDI), particularly in high fixed cost industries (e.g., oil exploration, petrochemicals), can create rents that an autocrat can use to fund the military. In doing so, these governments are able to retain the loyalty of a key domestic ally. Moreover, for many countries attracting foreign capital is part of a broader strategy of fostering economic growth.

I proceed as follows. The next section describes (briefly) the "varieties" of FDI and hones in how foreign investments in high fixed cost and capital-intensive sectors can create rents that a government can potentially benefit from. This discussion builds on the theoretical framework in Chapter 3 that government rents are an important source of political survival, particularly in nondemocracies. Sections 6.3 to 6.5 provide cross-national evidence to test the claim that FDI can strengthen nondemocratic rule by funding the military.

To identify a causal relationship, the empirical strategy in Section 6.2 describes a novel "hit or miss" strategy associated with the randomness of oil discoveries. In particular, the identification strategy exploits the "randomness"

---

[1] O'Donnell (1979); Wintrobe (1998); Bueno de Mesquita et al. (2003). O'Donnell (1979), for example, explains this tendency as a product of the inherent inability of autocratic regimes to develop mediations in their political systems between the various factions of the capital-owning classes and between capital and labor that are necessary to hold the system together when economic conditions deteriorate.

of oil discoveries to evaluate how *new* opportunities for rents lead to higher inflows of FDI *and* military spending in *autocracies*. The empirical strategy is a two-step process. I first exploit plausibly exogenous variation in oil discoveries to evaluate how *new* opportunities for rents *causes* higher inflows of FDI. I show that a "new" oil discovery (in year t) increases inflows of FDI in the subsequent year (i.e., year $t + 1$). I then show how this oil discovery–induced FDI inflow *causes higher* levels of military spending *in autocracies*, but not in democracies. The empirical sample includes fifty-nine oil-producing countries that receive inward FDI.

Yet, as scholars recognize, higher military spending may potentially embolden the military to overthrow the ruling government.[2] To discount this argument, I show that oil discovery–induced FDI does not heighten the likelihood of coups, civil wars, or accelerate the exit of dictators. The latter finding is consistent with the prediction from Chapter 3 that *in equilibrium*, if dictators are optimizing then a surge in *new* rents should *not* undermine their rule. These results – which are presented and discussed in Section 6.4 – show that random oil discoveries cause a surge in FDI that helps fund the military in autocracies but *does not* threaten an autocrat's political survival. Finally, Section 6.6 further traces the causal process by describing how attracting foreign capital played an integral part in sustaining General Suharto's dictatorial rule in Indonesia for over three decades. The case vignette highlights the importance of attracting foreign investment (particularly in the natural resource sectors) in the regime's governing strategy as a means to accelerate the country's industrialization and supply of patronage (rents) to key regime supporters. The case vignette also describes how outflows of foreign capital during the Asian Financial Crisis played an important role in the ultimate demise of the Suharto regime. Section 6.7 concludes.

6.1 THE POLITICS OF FDI

The Varieties of FDI

As conceptualized in Chapter 2, FDI entails the transfers of financial assets (and potentially other firm-specific assets, such as managerial and technological expertise) from a firm in a home country to a firm in a host country. The principal aim of the investment is to either establish business operations or acquire business assets, such as ownership or a controlling interest in the host country firm. As such, these foreign investments – like remittance income – do not directly accrue to a government's revenue base. However, unlike remittance income but like foreign aid, there are many "varieties" of FDI. These varieties can be differentiated, for example, based on the intended purpose of the investment (e.g., market-seeking, resource-seeking), its method (e.g., green

---

[2] Svolik (2012).

field investment, merger and acquisition), and sectoral allocation (e.g., in natural resources, retail), among others.[3] From an operational standpoint, FDI is frequently classified by its "type." For instance, horizontal FDI arises when a firm (investor) duplicates its home country–based activities in the host country. In contrast, vertical FDI occurs if a company invests in a foreign enterprise that plays the role of a supplier or a distributor. And in many instances, investors use FDI as a "platform" to export to a third country (e.g., investments by US firms in Japan as a means to export to China).

For host countries and their firms, inward FDI can be beneficial, as it can be a means of infusing financial capital, transferring technological knowledge across borders, generating economies of scale and scope, increasing trade linkages, and potentially generating "spillovers" to the broader economy.[4] Given these potential benefits, many governments actively court FDI, for example, by offering numerous forms of "incentives" to investors (e.g., tax holidays, subsidies, preferential tariffs), signing investment and trade treaties, and undertaking domestic economic and political reforms.[5] The willingness of governments to attract FDI to their countries, especially from multinational corporations (MNCs), has generated a large literature in economics and politics to better understand this "engagement."

Existing scholarship on this engagement can be categorized into three broad categories. One strand of this research examines why countries have become more open to FDI, for instance by analyzing public attitudes toward FDI and the sources of policies that attract FDI, such as lowering domestic regulations and signing investment treaties. A second strand considers how potential host country politics and policies influence the choice of countries in which firms

---

[3] The three most common motivations for FDI are market-seeking, resource-seeking, and efficiency-seeking. In an attempt to penetrate the local markets of host countries, market-seeking factors include the host country's market size, growth, and structure. Investments may also be resource-seeking in order to access raw materials, low-cost unskilled and skill labor, technological assets, and physical infrastructures (e.g., ports, highways) in the host country. Resource-seeking investments target specific sectors, such as natural resources (e.g., oil, mining), utilities, and transportation. Relatedly, FDI can be efficiency-seeking so as to benefit from factors that enable it to compete in international markets, such as greater (cheaper) access to natural resources, labor, other inputs, as well as transportation routes. In an effort to achieve these objectives, a foreign direct investor may gain a controlling interest in an enterprise in the host country through two types of strategies. The first strategy is for a home country company to set up new entities (e.g., factories, plants, stores) from the ground up in the host country. This method is called a green field investment. Companies such as Starbucks and McDonald's tend to use the green field approach when expanding overseas. The second strategy is through cross-border mergers and acquisitions that involve acquiring an existing foreign enterprise in the host country. This method is called a brown field investment.

[4] For an excellent treatment on the political economy of FDI, see Jensen et al. (2012).

[5] There are a wide range of incentives governments may offer to investors to attract FDI. Some of these include low corporate and tax rates, tax holidays, preferential tariffs, access to special economic zones, bonded warehouses, investment financial subsidies, free land or land subsidies, and infrastructure subsidies.

invest.[6] A third strand gauges the effect of FDI on governance, human rights, and inequality in host countries, often identifying differences across democracies and nondemocracies. The empirical analysis in this chapter falls squarely in this third strand. In the next subsection, I describe how FDI in sectors with fixed assets can provide opportunities for rent creation.

FDI and Rents

Chapter 3 argues that inward FDI can create opportunities for rents. This opportunity tends to be especially high in industries that require fixed assets and predominantly immobile investments.[7] Important features of fixed asset investments are the high initial capital requirements and the substantial sunk costs that act as entry barriers. These features influence market competition, and in particular create opportunities for monopoly or oligopoly rent extraction, especially in investment projects that involve state-owned enterprises, such as national oil companies and their subsidiaries. These subsidiaries include firms engaged in downstream refinement, distribution, and petrochemical products.

Why is this the case? By definition, expenditures on fixed assets inject large amounts of liquid capital into land, buildings, machinery, equipment, and technical expertise. For most firms, especially in developing countries, the existence of these high capital requirements creates entry barriers for potential entrants.[8] These entry costs are further heightened, as fixed asset investments typically incur relatively high sunk costs that are not recoverable and thus require substantial economies of scale in production to generate profit.[9] The confluence of these factors – large initial injection of capital and economies of scale in sunk costs – serve as barriers to potential entrants. These high entry barriers imply that only large firms can successfully enter; and in many developing countries, multinational corporations and state-owned enterprises are the only viable large firms. The presence of a few firms leads to market concentration which gives rise to opportunities for extracting monopoly or oligopoly rents. Thus, industries that require fixed asset investments (e.g., petroleum, mining, logging) enhance an MNC's incentive and ability to pursue monopoly or oligopoly rents. Moreover, if these MNCs partner with state-controlled firms, these rents are available to the government as well.

The oil industry embodies the quintessential high entry barrier sector with substantial amounts of FDI. Oil exploration and production are capital-intensive processes that require significant investments in fixed assets (e.g.,

---

[6] As Pandya (2016) notes this strand is well explored, as "the single most-studied research question about the politics of FDI in the last two decades concerns the types of political regimes MNCs prefer for the countries in which they invest." For an overview on the political economy of FDI, especially for the first and second strands of research, see Pandya (2016).
[7] Pinto and Zhu (2016).   [8] Bain (1956); Geroski (1995).   [9] Bain (1956); Harrigan (1981).

drilling wells, off-shore platforms, distribution pipes) well before a single barrel of oil is even pumped. From a production standpoint, the global oil industry – from exploration, production, and distribution – exhibits significant economies of scale and tends to be oligopolistic, primarily with large multinational corporations (e.g., Shell, British Petroleum) partnering with nation oil companies (e.g., Brazil's Petrobras, Saudi Arabia's Aramco) to generate substantial profits. It is not surprising therefore that FDI in oil discovery, refining, distributed, and affiliated industries (e.g., petrochemicals, energy distribution, construction) rank among the highest in the world. For example, between 2003 and 2015, around 22 percent of all green field investment globally was channeled into petroleum (i.e., discovery, refining), petrochemicals, and petroleum-related industries.[10] This share exceeds that of any other industry, such as electrical equipment, public utilities, and transportation. The prominence of FDI in the global oil industry thus offers a meaningful test of whether FDI can generate rents for a dictatorship to spend on its military.

Before turning to the empirics – which focuses on FDI associated primarily with the oil sector and potential upstream and downstream sectors – it is worth stressing that this conjecture could be generalized to other industries where a government has the potential to extract rents, such as logging, mining, energy distribution, infrastructure, transportation, and advanced technologies. That said, in the strictest sense, the causal estimates presented in this chapter apply to oil-producing countries that receive inward FDI.

6.2 HIT OR MISS: IDENTIFYING THE EFFECT OF OIL DISCOVERIES ON POLITICS

Identification Strategy

Evaluating the causal effect of oil wealth on inward FDI and political outcomes is challenging because oil exploration (and the resulting oil wealth) may be endogenous to a country's politics.[11] Several authoritarian regimes have, for example, actively pursued policies to attract foreign investment in its oil industry to legitimize and strengthen its rule, such as the "Triple Alliance" under Brazil's military dictatorship and General Suharto's "New Order" in Indonesia.[12] To mitigate the endogeneity of oil exploration with politics, I exploit the apparent *randomness* in oil exploration.

---

[10] Based on the author's calculations from the 2016 World Investment Report, Annex Table 20. A green field investment is a form of FDI where a parent company builds its operations in a foreign country from the ground up. bal oil industry, thusthe chapter.s a plausible test of whether FDI in a high fixed cost industr

[11] David and Wright (1997); Collier (2010).

[12] Relatedly, there is a large empirical literature documenting how regime type affects FDI flows. See for example, Jensen (2003), Li and Resnick (2003), and Asiedu and Lien (2011).

6.2 Hit or Miss: Identifying the Effect of Oil Discoveries on Politics

While oil exploration can be financially lucrative, it is inherently a hit-or-miss business. In practice, it is rare for the first exploratory borehole (also known as a wildcat) in a new area to succeed. For example, even with the most advanced technology, the success rate of exploration drilling is still less than 50 percent and was even lower for most of the twentieth century. According to the US Energy Information Administration, only 46 percent of exploratory wells found commercially viable quantities in 2003; this success rate was only 22 percent in 1973.[13] This randomness in the success of the oil exploration underlies my identification strategy.

To establish causality, I exploit changes in oil reserves over time due to the randomness in the success or failure oil exploration, *conditional* on the intensity of oil exploration attempts and other observables. My empirical approach is similar to those of Jones and Olken (2009) and Cotet and Tsui (2013).[14] The innovation in this chapter is to link foreign direct investment as a plausible channel through which oil discoveries – and subsequent opportunities for rents – affect military spending.

In an innovative paper, Jones and Olken (2009) evaluate the causal effect of assassination on various political outcomes, such as civil war, foreign aggression, institutional change, and the longevity of political leaders. To identify this effect, Jones and Olken employ the inherent randomness in whether an assassination attempt is successful or not.[15] Following their approach and Cotet and Tsui's (2013) extension to oil exploration, I estimate regressions of the form

$$Y_{it} = \beta DISCOVERY_{it} + \theta X_{it} + C_i + D_t + \varepsilon_{it} \qquad (6.1)$$

where Y_{it} is the outcome of interest for i in year t. Specifically, these outcomes are the change in military spending and net FDI inflows *before and after the exploration*. Thus, the estimating sample is restricted to the year before and after discovery attempts.[16] $DISCOVERY_{it}$ measures the value of new oil discoveries. It is the product of oil reserves and oil prices.

The key identifying assumption is that $DISCOVERY_{it}$ is exogenous (i.e., uncorrelated with ε_{it}) conditional on observables. Then $E[\varepsilon_{it}|DISCOVERY_{it}, X_{it}, C_i, D_t] = 0$ and β measures the average treatment effect of "new" oil revenues from successful new discoveries (relative to

---

[13] Statistic cited in Cotet and Tsui (2013).
[14] Cotet and Tsui (2013) examine the effect of oil discoveries on conflict. They do not examine the effect of oil discoveries on FDI.
[15] As they note, for example, President John F. Kennedy did not escape the bullet that killed him, even though it was fired from 265 feet away and the president was in a moving car. However, Ugandan President Idi Amin (from 1971–1979) did survive an attack in 1976 when a grenade bounced off of his chest and killed several bystanders.
[16] As a result, the estimating samples in the regression samples are smaller than those reported in Chapters 4 and 5. The analyses in those chapters are able to exploit the broader panel structure of country–year observations.

unsuccessful discoveries) on Y_{it}, after controlling for observables. A positive and significant coefficient on β implies that new oil wealth causes a short-run increase in inward FDI and military spending.

In equation (6.1), the observables include a measure of exploration intensity (i.e., the number of wildcat wells in an area where no area oil production exists). I include this variable because even if the likelihood of success is exogenous conditional on observables, the probability of successful oil exploration may be increasing in the number of attempts. I also control for lagged value of oil reserves in order to isolate the effect of "new" oil wealth (as result of a successful oil discovery) on Y_{it}. X_{it} also includes per capita income, economic growth, population, and democracy (all lagged by one year).

I control for country and decade fixed effects (C_i and D_t, respectively). Decade fixed effects are included to capture technological improvement in oil exploration over time. Of particular importance is the inclusion of country fixed effects that account for country-specific, time-invariant observable (e.g., colonial heritage, geography), and unobservable characteristics. With this fixed effects specification, β therefore gauges the within-country effect associated with new oil wealth on *changes* in military spending and net FDI inflows in the year before and after random (exogenous) oil discoveries.

Data

I use data from several sources. With respect to the dependent variables, data on defense spending are from the Stockholm International Peace Research Institute and net inflows of FDI are from the World Bank's World Development Indicators. Each dependent variable is the change from the year before $(t - 1)$ and after $(t + 1)$ an oil exploration. The Association for the Study of Peak Oil (ASPO) records data on oil discoveries. It contains a panel of oil discovery and production figures for the top sixty-two oil-producing countries over the 1930–2003 period. These data allow for the construction of a long panel of oil wealth, particularly by exploiting significant variation in oil wealth stemming from new (and even first) oil discoveries. Each country's oil revenues in any year are calculated by subtracting cumulative production from cumulative discovery.[17] The ASPO data set also provides information on oil exploration as represented by the number of wildcats drilled in any given year. Data on economic growth, income, and population are drawn from the World Bank's World Development Indicators, while the measure of democracy is from POLITY IV.

The full estimating sample includes fifty-nine oil-producing countries since 1971. These include ten countries in OPEC, as well as forty-nine that are not members of the oil cartel. The sample of countries spans the entire globe. The countries also vary in regime type over the sample period. There are thirty-six

[17] It is an unbalanced panel because some countries are not independent over the sample period.

countries that exhibit democratic governance: some for the entire sample period (e.g., Norway, India, United States) and others for some period (e.g., Russia, Mexico). There are thirty-nine countries that exhibit nondemocratic governance: 11.5 percent are monarchies, 15.6 percent are personalist dictatorships, 53.3 percent are single-party regimes, and 22.6 percent are governed by the military.

6.3 THE IMPACT OF OIL DISCOVERIES ON FDI AND MILITARY SPENDING

To establish a causal relationship from inward FDI on military spending associated with "random" oil discoveries, I proceed as follows. I first provide evidence that oil exploration is plausibly exogenous (i.e., uncorrelated with various observable characteristics). I then show that random oil discoveries – and the new oil wealth they generate – increase military spending in autocracies, but not in democracies. This finding is *not* due to potential outliers (e.g., countries in the Middle East), changes in other types of government spending, or a country's security environment. I then evaluate channels in Section 6.4. In this section, there are three key results. First, oil discoveries increase FDI in nondemocracies. Second, FDI is a channel through which oil discoveries lead to greater military spending in nondemocracies. Third, I discount other plausible channels that may be correlated with oil discoveries and foreign investment (e.g., economic growth, trade, sovereign borrowing) and show that oil wealth does not foster political instability.

Is Oil Exploration Random?

The key identification assumption for the main analysis is that *conditional* on drilling attempts (and other observables), "successful" oil discoveries are uncorrelated with the error term in equation (6.1). Basically, is success in oil exploration random? While this question cannot be tested directly, one approach is to investigate whether any observed variables predict successful oil exploration.

Cotet and Tsui (2013) analyzed the effect of oil rents on conflict. They found that that success in oil exploration is largely uncorrelated with various observable country characteristics that are theorized to correlate with political stability. Table 6.1 presents the mean value of a number of variables in the year prior to wildcat drilling and the associated results from two-sided t-tests for the equality of means between successful and unsuccessful wells.[18] A statistically insignificant "difference" implies that the variable under consideration (e.g., log GDP per capita) does not "predict" whether a wildcat well is successful or not.

---

[18] Table 6.1 replicates table 6 from Cotet and Tsui (2013).

TABLE 6.1 *Are successful and failed oil explorations similar?*

	Success	Failure	Difference
	(1)	(2)	(3)
Log oil wealth per capita	1.546	−2.462	4.008***
	(3.565)	(6.984)	(0.825)
Log GDP per capita	8.361	8.287	0.074
	(0.891)	(0.976)	(0.103)
Growth in GDP per capita (% annual)	2.233	2.152	0.081
	(5.804)	(7.110)	(0.326)
Log population	9.994	9.289	0.705***
	(1.598)	(1.480)	(0.193)
Log population density	3.258	3.522	−0.263
	(1.450)	(1.636)	(0.195)
Democracy	0.501	0.588	−0.086**
	(0.385)	(0.392)	(0.043)
Time-invariant characteristics			
Ethnic fractionalization	0.438	0.38	0.059*
	(0.253)	(0.270)	(0.033)
Religious fractionalization	0.387	0.423	−0.036
	(0.252)	(0.241)	(0.026)
Language fractionalization	0.3	0.282	0.018
	(0.269)	(0.240)	(0.032)
British legal origin	0.25	0.266	−0.016
	(0.433)	(0.442)	(0.049)
Wildcat	1.923	0.39	1.533
	(9.601)	(2.374)	(1.284)
No. of observations	2,042	613	2,655

Notes: The table replicates the analysis in Tsui and Cotet (2013, table 6). This table reports the means of each listed variable for the country–year prior to wildcat drilling leading to oil discovery (column 1) or failing to lead to oil discovery (column 2), with the standard deviation reported in parentheses. Column 3 reports the difference in means. The wildcat variable is divided by 100 to improve readability. The actual number of observations varies from variable to variable. *, **, *** = significant at 10%, 5%, and 1% respectively.

Table 6.1 illustrates that the two subsamples are "balanced" across several observable country characteristics including per capita wealth (log), economic growth, population density, the number of wildcats drilled, and several time-invariant country characteristics, such as religious and language fractionalization

6.3 Impact of Oil Discoveries on FDI and Military Spending

and legal origin. Differences in geography (i.e., log of percent mountainous terrain) and ethnic fractionalization are only marginally significant. A few observables are correlated with successful drilling: oil wealth per capita, population, and democracy. As Cotet and Tsui (2013) observe, although previous exploration seems to more successful in more populated and less democratic countries, the difference is rather small in the case of population and the difference is statistically insignificant with alternate measures of democracy, such as x-polity from Vreeland (2008). To mitigate any bias these correlations may introduce, *all* the regressions control for these observables.

Oil Discoveries and Military Spending

Having established that oil discoveries are plausibly exogenous, I now evaluate whether they *cause* a change in military spending. Panel A in Table 6.2 shows that oil discoveries increase subsequent military spending, particularly in autocracies. For the full sample of countries, column 1a shows that oil discoveries are positively correlated with military spending, but this effect is not statistically significant. This effect, however, seems to mask the differential effect in democracies and autocracies. To evaluate these differential effects across regime type, I employ a split sample approach in which a country with a POLITY score less than +7 is classified as a nondemocracy.[19]

In democracies, new discoveries tend to decrease subsequent military spending, although this effect is not statistically significant (column 2a). In contrast, oil discoveries *cause* a large and statistically significant increase in military spending in autocracies (column 3a). The coefficient estimate (= 0.97) implies that a 1 standard deviation increase in new oil wealth raises military spending by 6.8 percentage points (of GDP). This is substantively important, as it is *double* the average annual change in military spending in autocracies.

While the success of oil explorations is random, skeptics may worry that the derivation of oil revenue from oil prices may not be (recall, value of oil discoveries = new oil reserves × current world oil price). To allay this concern, I instrument for *oil prices* in two ways. The first identification strategy follows that of Ramsay (2011) and exploits out-of-region natural disaster damage for oil-producing nations as an instrument for oil prices. Out-of-region natural disasters curtail oil production and raise world oil prices. The instrument comprises the out-of-region disaster, oil reserves (both in log units), and the product of these two variables (column 4a).[20] The second identification strategy

---

[19] The +7 POLITY cutoff follows convention among scholars in political science. This split sample approach follows the strategy in Jones and Olken (2009) and Cotet and Tsui (2013).

[20] I consider five types of natural disasters (earthquake, volcano, mass movement, storms, and floods) in oil-producing countries by aggregating the value of disaster damage in six regions. Out-of-region disaster is measured as the value of all disaster minus the value of own region damages.

TABLE 6.2 *New oil discoveries increase military spending in nondemocracies*

Dependent variable:	A: Change in defense burden				
Sample:	All countries	Democracies	Autocracies	Autocracies	Autocracies
Estimation:	OLS (1a)	OLS (2a)	OLS (3a)	2SLS (4a)	2SLS (5a)
Log value of oil discovery per capita	0.362 (0.252)	−0.259 (0.238)	0.971 (0.333)***	0.972 (0.305)***	0.794 (0.289)***
Instrument				Natural disasters	Expected discovery
First-stage F-statistic				8.11×10^4	2.21×10^4
No. of observations	628	392	236	231	196
No. of countries	57	35	33	28	25

Dependent variable:	B: Change in components of military spending		
	Arms import	Personnel	Per person military expenditure
Sample:	Autocracies	Autocracies	Autocracies
Estimation:	OLS (1b)	OLS (2b)	OLS (3b)
Log value of oil discovery per capita	−63.55 (779.271)	152.469 (94.540)^	12.237 (5.570)**
No. of observations	168	221	204
No. of countries	25	31	30

Notes: All coefficients (and standard errors) are multiplied by 100 to improve readability. In all columns, the samples include all country–year observations with at least one wildcat drilling. Estimation via ordinary least squares (OLS) in columns 1a to 3a, 1b to 3b. Estimation via two-stage least squares (2SLS) in columns 4a and 5a. In panel A, the dependent variable is the change in military spending (% GDP) 1 year before and 1 year after wildcat drilling. In panel B, the dependent variables are the change in arm imports (US$, millions), military personnel (thousands), and military expenditures per military official (US$, thousands) 1 year before and 1 year after wildcat drilling. In column 4a, I use log(out-of-region disaster), log(oil discoveries per capita), and their product as instruments. In column 5a, I use unexpected oil discovery as an instrument. All the t-statistics associated with the instruments in the first-stage regressions are significant at the 1% level. The Hansen J-test of overidentification fails to reject the null that the instruments are valid, i.e., not correlated with the error term at conventional significance levels. All regressions control for the number of wildcats, country and decade fixed effects, and 1-year lags of log GDP per capita, growth in GDP per capita, log population, democracy, log value of oil reserves. These coefficients and a constant are not reported. Robust standard errors clustered at the country level are reported in parentheses. ^, *, **, *** = significant at 11%, 10%, 5%, and 1% respectively.

6.3 Impact of Oil Discoveries on FDI and Military Spending

uses the unexpected component of an oil discovery as an instrumental variable for the value of oil discoveries. In particular, the spatial correlation of oil endowments implies that recent oil discoveries in neighboring countries generate plausible estimates for the unexpected component in oil discoveries that is exogenous to local economic and political conditions.[21] I use the residual between actual discoveries and expected discoveries to obtain the measure of unexpected discoveries (column 5a). With each instrument, new oil wealth causes an increase in military spending in autocracies (columns 4a and 5a).[22]

Composition of Military Spending

This surge in military spending seems to benefit military personnel.[23] Panel B in Table 6.2 evaluates the effect of new oil discoveries on two important components of military spending in autocracies: arms imports and military personnel. The former captures an important source of military hardware in dictatorships.[24] Column 1b shows that oil discoveries, surprisingly, lead to a decrease in arm imports, although this effect is not statistically significant. In contrast, oil discoveries *positively* affect the well-being of military personnel in autocracies. The coefficient estimate in column 1b (coefficient = 152.5, $p = 0.11$) implies that a 1 standard deviation increase in new oil wealth expands the military by around 10,700 individuals.

Moreover, new oil wealth improves the welfare of the "typical" military official. The positive and statistically significant coefficient in column 1c implies that new oil wealth increases average expenditures on each military official, some of which is likely to be in the form of higher wages and more generous benefits (e.g., better health care, subsidized housing). Together, the positive

---

[21] Cotet and Tsui (2013). The instrument is calculated as follows. For each country–year with a positive number of exploration attempts, I calculate the average discovery per wildcat for *other* countries in the same geographical region over the previous 5 years. I then multiply the average discovery per wildcat by the number of wildcats drilled in the country of interest to construct a variable used to forecast the expected discoveries. Following Card (1990), I generate expected discovery as the product of regional discoveries per wildcat and the number of wildcats in that country. I then regress actual discoveries on the measure of expected discoveries to obtain unexpected discoveries as the residuals.

[22] Each instrument is "strong," as the F-statistic exceeds the threshold for weak instruments (i.e., typically a value of 10 for a single instrumental variable). And for completeness, using the same instruments for oil prices, I verify that new oil wealth does not affect military spending in *democracies*. With the "natural disasters" instrument, the coefficient on oil wealth is –0.26 with a corresponding standard error of 0.226 ($p = 0.25$). With the "unexpected discovery" instrument, the coefficient is –0.19 with a corresponding standard error 0.204 ($p = 0.36$).

[23] Data on these components of military spending are less comprehensive than those of total military spending. Thus, I use total military expenditures as my preferred measure in this chapter.

[24] Oil-producing autocracies are not the leading manufacturers of weaponry. Rather, the leading arms manufacturers and exporters tend to be advanced democracies (e.g., the United States, United Kingdom, France). Some exceptions are Russia and China.

coefficients in columns 2b and 3b imply new oil wealth raises the welfare of military officials, which is likely to maintain their loyalty to the regime.

The main finding is robust to various concerns.[25] First, the results are robust to excluding potential outliers, such as countries that belong to OPEC and those in the Middle East and North Africa (e.g., Saudi Arabia), countries that also tend to be nondemocratic. Second, oil discoveries do not exhibit an "income effect" for other aspects of government spending. Oil discoveries do not exhibit statistically significant effects on total government spending or on nonmilitary government expenditures.[26] Oil discoveries cause a robust increase in military spending in dictatorships only.

Third, it is plausible that oil exploration intensifies security competition in the country and/or region surrounding the oil discovery; and this heightened risk of military disputes drives the surge in military spending following an oil discovery.[27] While plausible, this does not seem to be the case. Columns 1 to 3 in Table 6.3 show that controlling for the incidence of any militarized interstate dispute (MID) involving the country with the oil discovery or any conflict involving its regional neighbors in the current year or past 5 years does not attenuate the robust positive effect of new oil wealth on military spending. Furthermore, columns 4 to 7 show that new oil wealth does not spark any military disputes involving the country with the oil discovery or among countries in its region.[28] Together, these findings imply the increase in military spending following oil discoveries is unlikely to stem from heightened security competition associated with oil exploration.

6.4 EVALUATING CHANNELS

Oil Discoveries Attract Foreign Investment

Having demonstrated that oil discoveries cause a short-run increase in military spending, I now examine whether inward FDI is a viable channel. To do so, I first evaluate whether new oil wealth also causes a surge in inward FDI. The results are reported in Table 6.4, panel A. For the full sample of countries, column 1a shows that oil discoveries lead to a slight bump in FDI inflows the year after, but this effect is not statistically significant. As with military spending, this seems to mask the differential effects by regime type. In

---

[25] These results are included in the book's replication package.
[26] These results show the coefficient estimates associated with nonmilitary spending are smaller in magnitude than those on total government spending. This suggests that oil-induced *military spending* accounts for significant amounts of changes in government spending.
[27] Data on militarized international disputes is from the Correlates of War (Sarkees and Wayman 2010).
[28] While oil discoveries are positively correlated with the onset of an MID involving the country with the oil discovery, the effect is not statistically significant. Moreover, new oil wealth tends to lower the propensity for interstate conflict in the region.

6.4 Evaluating Channels

TABLE 6.3 *Oil discoveries do not spur security competition*

Dependent variable:	Change in *dependent variable* in autocracies						
	Defense burden			Ongoing MID		MID in region	
Estimation:	OLS	2SLS	2SLS	OLS	OLS	OLS	OLS
	(1)	(2)	(3)	(4)	(5)	(6)	(7)
Log value of oil discovery per capita	1.015 (0.354)***	1.015 (0.320)***	0.881 (0.319)***	0.409 (0.247)		−0.025 (0.110)	
Log FDI					−0.007 (0.011)		0.002 (0.004)
Ongoing MID	−0.057 (0.035)	−0.057 (0.031)*	−0.072 (0.037)*				
Any MID in past 5 years	0.015 (0.083)	0.015 (0.073)	0.052 (0.073)				
Ongoing MID in region	−0.032 (0.181)	−0.032 (0.160)	−0.074 (0.203)				
Any MID in past 5 years in region	−0.320 (0.296)	−0.32 (0.262)	−0.424 (0.298)				
Instrument		Natural disasters	Expected discovery				
First stage F-statistic		8.7×10^4	2.9×10^4				
No. of observations	236	231	196	1,347	513	1,347	513
No. of countries	33	28	25	50	39	50	39

Notes: All coefficients (and standard errors) are multiplied by 100 to improve readability. In all columns, the samples include all country–year observations with at least one wildcat drilling for *autocracies*. The dependent variable in columns 1 to 3 is the change in a country's defense burden from 1 year before and 1 year after wildcat drilling. In columns 4 and 5, the dependent variable is the change in the incidence of a MID involving the country with the oil discovery from 1 year before and 1 year after a wildcat drilling. In columns 6 and 7, the dependent variable is the change in the incidence of a MID involving any country in the same region with an oil discovery from 1 year before and 1 year after a wildcat drilling. In column 2, I use log(out-of-region disaster), log(oil discoveries per capita) and their product as instruments for log value of oil discoveries. In column 3, I use unexpected oil discovery as an instrument for log value of oil discoveries. All the *t*-statistics associated with the instruments in the first-stage regressions are significant at the 1% level. The Hansen J-test of overidentification fails to reject the null that the instruments are valid, i.e., not correlated with the error term at conventional significance levels. All regressions control for the number of wildcats, country and decade fixed effects, and 1-year lags of log GDP per capita, growth in GDP per capita, log population, democracy, and log value of oil reserves. These coefficients and a constant are not reported. Robust standard errors clustered at the country level are reported in parentheses. *, **, *** = significant at 10%, 5%, and 1% respectively. 2SLS, two-stage least squares; FDI, foreign direct investment; MID, militarized interstate dispute; OLS, ordinary least squares.

TABLE 6.4 *Oil discoveries increase FDI and military spending*

Dependent variable:	A: Change in FDI				
Sample:	All countries	Democracies	Autocracies	Autocracies	Autocracies
Estimation:	OLS	OLS	OLS	2SLS	2SLS
	(1a)	(2a)	(3a)	(4a)	(5a)
Log value of oil Discoveries	0.889 (0.693)	−0.009 (0.624)	2.941 (1.721)*	3.198 (1.745)*	4.159 (1.950)**
Instrument				Natural disasters	Expected discovery
First-stage F-statistic				1.7×10^4	1.3×10^3
No. of observations	1,105	682	423	420	345
No. of countries	59	36	39	36	33

Dependent variable:	B: Change in defense burden					
	With an increase in FDI			With NO increase in FDI		
Sample:	Autocracies	Autocracies	Autocracies	Autocracies	Autocracies	Autocracies
Estimation:	OLS	2SLS	2SLS	OLS	2SLS	2SLS
	(1b)	(2b)	(3b)	(4b)	(5b)	(6b)
Log value of oil discoveries	0.811 (0.386)**	0.803 (0.347)**	0.669 (0.359)*	0.627 (1.1724)	0.655 (1.145)	−0.955 (0.592)
Instrument		Natural disasters	Expected Discovery		Natural disasters	Expected discovery
First stage F-statistic		4.60×10^4	9.85×10^4		7.9×10^3	2.47×10^3
No. of observations	176	170	146	60	49	36
No. of countries	32	26	22	24	13	9

Notes: All coefficients (and standard errors) are multiplied by 100 to improve readability. In all columns, the samples include all country–year observations with at least one wildcat drilling. In Panels A and B the dependent variables are the change in log FDI (2010 US$) and defense burden 1 year before and 1 year after wildcat drilling. In columns 1b to 3b, the sample is restricted to country–year observations in which there is an increase in inward FDI from the previous year. In columns 4b to 6b, the sample is restricted to country–year observations in which there was no an increase in inward FDI from the previous year. In columns 4a, 2b, and 5b, I use log(out-of-region disaster), log(oil discoveries per capita), and their product as instruments. In columns 5a, 3b, and 6b, I use unexpected oil discovery as an instrument. All the *t*-statistics associated with the instruments in the first-stage regressions are significant at the 1% level. The Hansen J-test of overidentification fails to reject the null that the instruments are valid, i.e., not correlated with the error term at conventional significance levels. All regressions control for the number of wildcats, country and decade fixed effects, and 1-year lags of log GDP per capita, growth in GDP per capita, log population, democracy, log value of oil reserves. These coefficients and a constant are not reported. Robust standard errors clustered by country are reported in parentheses. *, **, *** = significant at 10%, 5%, and 1% respectively.

6.4 Evaluating Channels

democracies, new discoveries have no effect on changes in inward FDI, as the coefficient is basically nil (column 2a). In contrast, oil discoveries generate a large and statistically significant surge in inward FDI in autocracies (column 3a). The coefficient estimate (= 2.94) implies that a 1 standard deviation increase in oil discoveries attracts about $112 million in new FDI. This effect is substantively important, as it corresponds to a staggering 109 percent increase over the average annual change in FDI in these nondemocracies.

Military Spending and FDI

These results do not necessarily imply that oil discoveries cause an increase in military spending in autocracies via an increase in FDI in these countries. To better establish this causal pathway, I evaluate the effect of oil discoveries on military spending in autocracies that accompanies an increase in FDI from the previous year. The results are presented in panel B of Table 6.4.[29] For this sample, oil discoveries cause a significant increase in military spending in nondemocracies. The OLS and 2SLS estimates in columns 1b to 3b are slightly smaller in magnitude than those associated with autocracies that do *not* condition on FDI in Table 6.3, columns 3a to 5a. These smaller coefficient estimates are expected because these specifications condition on an increase in inward FDI.

Finally, for completeness, columns 4b to 6b show that oil discoveries *do not* increase military spending in autocracies when there is *no* increase in FDI inflows. It is worth noting that this country–year sample of autocracies (columns 4b to 6b) do not differ on observable characteristics from those that experience an increase in FDI (columns 1b to 3b). A statistical comparison of group means show there is no difference in the level of economic development; growth; value of oil existing oil reserves; number of wildcat drills; "quality" of autocracy (e.g., POLITY score); and various measures of political stability, such as the incidence of coups and civil war and the probability of leader "exit." These similarities, particularly for the measures of political (in)stability, imply the sample of countries that experience an increase in FDI (columns 1b to 3b) are *not* more stable autocracies that may be in a stronger position to exploit newly discovered oil fields, attract FDI, and thus fuel military spending. Thus, the null findings in columns 4b to 6b are reassuring and in conjunction with the positive and statistically significant coefficient estimates in columns 1b to 3b imply that oil discoveries cause an increase in military spending in autocracies *only* when accompanied with an increase in FDI.

Discounting Other Channels

The results in Table 6.4 show that oil discoveries allow dictators to increase military spending in years with a concurrent increase in FDI. This is consistent

---

[29] This reduces the number of observations in the full sample of autocracies (from columns 3 to 5 in Table 6.3) by about a third.

TABLE 6.5 *Discounting other channels*

Dependent variable:	Change in *dependent variable*, in autocracies						
					Defense burden		
	Growth	Imports	Exports	Reserves	*With an increase in FDI*		
Estimation:	OLS	OLS	OLS	OLS	OLS	2SLS	2SLS
	(1)	(2)	(3)	(4)	(5)	(6)	(7)
Log value of oil discovery per capita	0.032	−0.351	−0.368	−2.185	1.109	1.115	1.22
	(0.055)	(0.314)	(0.213)*	(6.236)	(0.592)*	(0.500)**	(0.443)***
Additional controls							
Log exports					−0.675	−0.676	−0.745
					(0.523)	(0.433)	(0.420)*
Log imports					0.399	0.399	0.451
					(0.244)	(0.201)**	(0.193)**
Reserves					0.010	0.010	0.019
					(0.014)	(0.011)	(0.010)
Instrument						Natural disasters	Expected discovery
First stage F-statistic						2.2×10^4	9.6×10^3
No. of observations	513	380	352	336	117	114	95
No. of countries	39	32	32	30	22	19	15

Notes: All coefficients (and standard errors) are multiplied by 100 to improve readability. In all columns, the samples include all country–year observations with at least one wildcat drilling for *autocracies*. In columns 1 to 4, the dependent variables are the change in economic growth, imports, exports, and total reserves (respectively) from 1 year before and 1 year after wildcat drilling. In columns 4 to 6, the dependent variable is change in defense burden from 1 year before and 1 year after wildcat drilling where FDI increased from the previous year. In column 6, I use log(out-of-region disaster), log(oil discoveries per capita), and their product as instruments for log value of oil discoveries. In column 7, I use unexpected oil discovery as an instrument for log value of oil discoveries. All the *t*-statistics associated with the instruments in the first-stage regressions are significant at the 1% level. The Hansen J-test of overidentification fails to reject the null that the instruments are valid, i.e., not correlated with the error term at conventional significance levels. All regressions control for the number of wildcats, country and decade fixed effects, and 1-year lags of log GDP per capita, growth in GDP per capita, log population, democracy, and log value of oil reserves. These coefficients and a constant are not reported. Columns 4 to 6 also control for log exports (2010 US$) and log imports (2010 US$). Robust standard errors clustered by country are reported in parentheses. *, **, *** = significant at 10%, 5%, and 1% respectively.

6.4 Evaluating Channels

with the chapter's main conjectures that (1) oil discoveries attract foreign investments, particularly in nondemocracies, which in turn (2) creates rents that a dictatorship can spend on its military. Of course, in addition to the FDI-induced military spending channel, it is plausible that oil discoveries may affect other economic variables that may benefit an autocrat (e.g., through higher international trade, higher economic growth, and sovereign borrowing). While the existence of these other pathways does not necessarily nullify the core findings in Tables 6.2 and 6.4 linking FDI to greater military spending in autocracies, it is important to evaluate whether greater trade, sovereign borrowing, and economic growth may also augur well for an incumbent. I do so in two ways.

First, I use the hit-or-miss specification to evaluate whether oil discoveries lead to an *increase* in economic growth, trade flows, and sovereign borrowing. These specifications are similar to those in Tables 6.2 and 6.4, except that the dependent variable is the change in economic growth, trade flows, and total reserves from one year before and one year after wildcat drilling. Table 6.5 reports the results across a sample of nondemocracies. In column 1, new oil discoveries are not a robust determinant of subsequently higher rates of economic growth. While the coefficient is positive (= 0.032), it is not statistically significant. Turning to international trade, columns 2 and 3 examine the effect of new oil discoveries on changes in imports and exports, separately. In both specifications, new discoveries do not increase a country's imports or exports and in particular they tend to decrease exports. As noted in Section 6.1, oil discoveries might also "signal" favorable future growth prospects. This, in turn, may reduce the costs of sovereign borrowing and increase a government's total reserves to potentially finance the military. While plausible, the evidence does not substantiate this channel. Column 4 shows that oil discoveries are negatively associated with changes in total reserves. Oil discoveries are also *not* a robust determinant of changes in external (sovereign) debt, portfolio investment, and foreign aid. Together, columns 1 to 4 show that oil discoveries do not subsequently increase economic growth, trade, or other types of foreign capital (e.g., reserves) – economic factors that could potentially benefit an autocrat.[30]

Second, while oil discoveries do not seem to increase economic growth, levels of trade, or access to foreign borrowing, it is plausible that these factors might attenuate the effect of oil discovery–induced FDI inflows on military spending. While the baseline results in Tables 6.2 and 6.4 control for economic growth, I supplement this analysis by also controlling for each country's imports, exports, and total reserves. Columns 5 to 7 show that controlling for these additional factors does not diminish the statistically significant positive effect of oil discoveries on military spending in countries that experience an increase in FDI. These results further confirm the finding in Table 6.2 that oil discovery induced FDI can generate rents to fund the military.

---

[30] Similar, null results are present in democracies.

TABLE 6.6 *Oil discoveries are not politically destabilizing*

Dependent variable:	A: Change in incidence of civil war					
Sample:	All countries	Democracies	Autocracies	All countries	Democracies	Autocracies
				With an increase in FDI		
	(1a)	(2a)	(3a)	(4a)	(5a)	(6a)
Log value of oil discoveries	0.079	0.238	−0.163	0	−0.12	0.412
	(0.210)	(0.309)	(0.302)	(0.132)	(0.240)	(0.385)
No. of observations	628	392	236	433	257	176
No. of countries	57	35	33	57	34	32

Dependent variable:	B: Change in coup incidence					
Sample:	All countries	Democracies	Autocracies	All countries	Democracies	Autocracies
				With an increase in FDI		
	(1b)	(2b)	(3b)	(4b)	(5b)	(6b)
Log value of oil discoveries	−0.241	−0.075	−0.449	−0.01	0.053	−0.106
	(0.334)	(0.095)	(0.662)	(0.453)	(0.232)	(0.612)
No. of observations	628	392	236	433	257	176
No. of countries	57	35	33	57	34	32

Dependent variable:	C: Change in *type of exit* in autocracies					
Type of exit:	All types	Regular	Irregular	All types	Regular	Irregular
				With an increase in FDI		
	(1c)	(2c)	(3c)	(4c)	(5c)	(6c)
Log value of oil discoveries	−0.256	−0.163	−0.138	−0.419	−0.387	−0.294
	(0.327)	(0.282)	(0.154)	(0.411)	(0.459)	(0.233)
No. of observations	235	235	235	175	175	175
No. of countries	32	32	32	31	31	31

Notes: Estimation via ordinary least squares. All coefficients (and standard errors) are multiplied by 100 to improve readability. All regressions control for the number of wildcats, country and decade fixed effects, and 1 year lags of log GDP per capita, growth in GDP per capita, log population, democracy, log value of oil reserves. These coefficients and a constant are not reported. In panel C, all the specifications control for leader duration polynomial. Robust standard errors clustered by country are reported in parentheses. *, **, *** = significant at 10%, 5%, and 1% respectively.

6.5 POLITICAL SURVIVAL

If dictators are optimizing their behavior as theorized in this book, then in equilibrium, they should optimally allocate their revenues to preserve their rule. Empirically, this means that the allocation of *additional* rents (e.g., new wealth from oil discoveries) should *not* threaten their political longevity. With this interpretation in mind, in this section, I provide evidence that new oil wealth – and the inward FDI that accompanies it – does *not* foster political instability, particularly in dictatorships. I estimate hit-or-miss specifications with changes in various measures of political stability as the dependent variable across samples split by regime type and in autocracies with an increase in FDI (after an oil discovery). Table 6.6 presents these *null* findings of oil discoveries on political instability.

Civil Conflict. While additional defense spending can help shore up a dictator's support from the military, it may in the process heighten the likelihood of civil conflict, by increasing both the state's repressive capacity and the value of capturing the "state prize" by opposition groups.[31] To investigate this concern, I evaluate whether oil discoveries lead to civil conflict as a robustness check. These specifications are similar to those in Table 6.2, except that the dependent variable is the change in the incidence of civil war from 1 year before and 1 year after wildcat drilling.[32] The results are presented in Table 6.6, panel A. Across the full sample of countries (column 1a), there is a slight positive but statistically insignificant relationship between the value of oil discoveries and conflict. There is no effect when there is also an increase in FDI (column 4a). In separate samples of democracies and nondemocracies, new oil wealth does not trigger civil conflict (columns 2a, 3a, 5a, and 6a). In *none* of these specifications is the coefficient on the value of new oil discoveries statistically significant.

Coup Propensity. Of course, it is plausible that by spending more on the military, an autocrat may embolden the military to mount a coup. This is a legitimate concern, as military coups are the leading cause of a dictator's removal from power.[33] The positive coefficients in columns 1b and 1c in Table 6.2 suggest that military personnel benefit from new oil wealth in autocracies and are therefore less likely to overthrow the incumbent government. Nevertheless, to further investigate this concern, I evaluate whether oil discoveries lead to a coup attempt as a robustness check. The results are presented in panel B of Table 6.6. Across the full sample of countries as well as for split samples of democracies and autocracies, there tends to be a negative (but statistically insignificant) relationship between oil discoveries and coup propensity. These null findings imply that oil discoveries do not cause coups.

Leader Survival. Finally, I investigate whether new oil wealth threatens the political survival of autocrats through "regular" (e.g., elections) or

---

[31] Besley and Persson (2010). [32] The measure of civil war status is from the Correlates of War.
[33] Svolik (2012).

"irregular" (e.g., coups) means using data from *Archigos*.[34] To do so, I employ a hit-or-miss approach and following Carter and Signorino (2010) control for a leader's incumbency with a polynomial of "temporal dependency." In these specifications, which are presented in panel C of Table 6.6, new oil wealth does not accelerate the exit of dictators through either regular or irregular means. Rather, new oil wealth tends to *lower* the likelihood of leader exit, but the effect is not precisely estimated.

On balance, the null findings associated with civil war, coup propensity, and leader exit, especially in autocracies – and those with a concurrent increase in FDI – are reassuring and ameliorate concerns that new oil discoveries make countries more politically unstable.

To summarize, the results in Tables 6.2 to 6.4 show that oil discoveries cause an increase in military spending and inward FDI in autocracies (Tables 6.2 and 6.4, respectively). And in autocracies, oil discoveries – accompanying an increase in FDI – cause dictatorships to spend more on their militaries (Table 6.4, panel B). This finding is not associated with other channels that are correlated (potentially) with FDI and oil discoveries (Table 6.5) and new oil wealth is not politically destabilizing, especially in dictatorships (Table 6.6).

6.6 FOREIGN INVESTMENTS IN SUHARTO'S "NEW ORDER"

This dynamic of rent creation between MNCs and governments has frequently played an important role in the development strategies of authoritarian regimes, especially for military regimes determined to industrialize their countries. In this section, I describe how General Suharto's regime actively courted foreign investment to legitimize its rule and expand the rents available to its key supporters. Yet, when these rents vanished during the Asian financial crisis, the regime lost its legitimacy and collapsed.

Foreign Capital in Indonesia's "New Order"

In 1949, the Dutch formally recognized Indonesian independence. The new republican government, however, inherited a commodity-based export economy in serious decline. In an effort to revitalize the economy and spur industrialization, the government formulated an industrial plan with a significant degree of private participation and push for investment. Despite this stated intention, by the late 1950s private and particularly indigenous capital proved incapable of effectively filling the required investment gap: if Indonesia intended to industrialize, the state had to take the lead.[35] However, political fragmentation hindered the adoption of a coherent state-led

---

[34] Irregular exits are more prevalent in nondemocracies. Data on leadership tenure and exits are from *Archigos* (Goemans et al. 2009).
[35] Thomas and Panglaykim (1973), 56–59.

industrialization policy. By the 1950s, weak economic conditions coupled with political disorder culminated in a political revolution from above, led by Sukarno with the support of the military. The new regime – authoritarian in governance, populist in style, and corporatist in structure – was short lived. Continuing economic crisis and political conflict led to a military takeover in 1965 (led by General Suharto), which effectively eliminated both Sukarno and the growing Communist Party (PKI) from power.

According to Robison (1988), the military takeover resolved two crises. First it settled the power struggle between the military and the PKI, the two cohesive political forces that had emerged from the previous period of political chaos. Second, the military takeover "reconstituted the path of capitalist development by bringing back international capital, the crucial and hitherto missing ingredient needed to clear up the crisis of investment and debt that underlay the country's economic problems."[36] Ideologues of the resulting "New Order" legitimized the regime's authoritarian rule as a necessary step in fostering objective, scientific, and decisive policymaking in order to build an industrial base for the economy and to provide the preconditions for future democratic governance.[37] At the outset of the New Order's ambitious economic and industrial policy, Suharto appointed a team of technocrats to revive the moribund economy, tackle the fiscal crisis, and enable investment to resume again.

In particular, rescheduling Indonesia's debts and negotiating new loans required assistance from international creditors (i.e., foreign governments and investors). In return for foreign capital, Indonesia's new leaders agreed to reduce the government's involvement in state-led industrialization and to encourage private investment in the consumer goods and agricultural sectors. Furthermore, the government agreed to draft a new foreign investment law that provided relief from taxation and other incentives to attract foreign investment.[38]

As a result, the New Order created a climate attractive to foreign capital. Initially, much of this foreign capital targeted Indonesia's oil sector and the government actively granted international corporations the right to invest in petrochemicals, fertilizers, steel, and metal engineering. The government, however, regulated foreign investment in various "selected" sectors in large part to protect domestic producers from foreign competition so as to develop home-grown infant industries.[39] Buoyant oil prices allowed the regime to channel oil revenues (which largely accrued to the government) into domestic

---

[36] Robison (1988), 60.   [37] Ward (1973).
[38] Thomas and Panglaykim (1973), 100–144. These new laws did not mark a move toward free market policies per se, but toward import substitution industrialization behind substantial tariff barriers in the hope and with the intention that private domestic and foreign capital would replace state corporations.
[39] Robison (1988).

investment. However, as the world price of oil tanked in the 1980s, domestic investment fell sharply. To fill this capital shortfall, the government greatly liberalized its FDI policy, thus allowing foreign capital into other sectors of the Indonesian economy.

Attracting foreign capital helped legitimize authoritarian rule in Indonesia in two distinct ways: by creating opportunities for patronage and by helping legitimize the government among the broader population. First, the regime's overall strategy to raise capital from both domestic and foreign sources created rents and potential channels of political patronage, especially for the military. The ability of the regime to strategically allocate funds (emanating primarily from oil revenues), "guide" foreign capital, and extend state protection afforded privileged access to the state bank credit, forestry concessions, trade and manufacturing monopolies, official distributorship of basic foodstuffs, and state contracts for supply and construction. The majority of these new corporate groups were Chinese-owned and their directors generally had previous associations with the military, acting as financiers for army commands and individual generals, and de facto managers for corporations owned by the military.[40]

The regime's commitment to attracting foreign capital also benefited government bureaucrats and individuals within Suharto's family and inner circle. For instance, Robison (1988, 64) observes that "the ideological commitment to industrial deepening also enhanced the financial base, economic power, and patronage of the major bureaucrats and state managers of capital because it was they who appropriated the authority to direct investment, allocate contracts, and grant import, monopolies." Access to foreign capital and its associated technology and knowhow helped enrich Suharto's friends and family in advanced and highly profitable sectors. According to Robison (1988, 64–65): "Led by Liem Sioe Liong, Willem Soerjadjaja, Tan Siong Ke, Bob Hasan, Hendra Rahardja, Ciputra, and various Suharto family interests including the Bimantara group, these corporate groups extended their industrial activities – invariably in joint ventures with foreign partners – into auto manufacture, tires, batteries, components, cement, electronics, into the engineering sectors that grew around the oil industry, and into the manufacture of steel."

In addition to creating opportunities for patronage, foreign capital fostered broader economic gains, further legitimizing the regime. The empirical literature documents consistent economic benefits from FDI in Indonesia.[41] Foreign firms bring in new production processes, start producing new goods (that benefit the entire country), and tend to be more productive than local firms.[42] These productivity gains translate into higher wages. Indeed, foreign-owned firms in Indonesia pay higher wages than domestically owned establishments, and pay higher wage premiums with a higher level of education.[43] In addition to raising wages, FDI fostered growth in Indonesia.

[40] Robison (1986). [41] Lipsey and Sjoholm (2010). [42] Takii and Ramsetter (2005).
[43] Lipsey and Sjoholm (2006).

Lipsey et al. (2010) find that foreign firms have relatively high growth in employment and foreign acquisitions of domestic firms increase in employment. Moreover, FDI in Indonesia has generated positive spillovers: local firms benefit from the presence of foreign firms or joint ventures between local and foreign firms within the same industry or region.[44] In particular, local firms often experience gains in productivity that translate to higher wages.[45]

Capital Outflows and Authoritarian Collapse

The political economy of foreign capital in Indonesia illustrates that inflows of international investment in rent-creating sectors can help legitimize and sustain authoritarian rule. Conversely, the withdrawal of foreign capital can be politically destabilizing. Suharto's response to the Asian Financial Crisis (1997–1999) provides stark evidence that outflows of foreign capital can accelerate the collapse of an authoritarian regime. In the years prior to the financial crisis, FDI inflows grew a moderate pace, reaching about 2.7 percent of GDP in 1996. However, by the second half of 1997, Indonesia became the country hardest hit by the economic crisis. By the end of the crisis, General Suharto had resigned, the policies of the New Order terminated, and Indonesia began a transition to more democratic governance.

While there are several accounts that strive to explain the demise of Suharto's regime, a leading narrative is the government's choice of macroeconomic policy response and its subsequent effects (negative) on the economy and political stability.[46] In response to the "twin" crises entailing simultaneous currency and banking crises, a government can tighten macroeconomic policy to counter rapid currency depreciation; or it can restrict capital flows and fix the exchange rate to make expansionary macroeconomic policy feasible; or it can adopt mixed policy measures. In contrasting the policy responses of the

---

[44] Temenggung (2006); Blalock and Gertler (2008).
[45] For instance, productivity spillovers include: new technologies and knowledge that are made available for domestic firms and increased competition. For wages, the positive effect of FDI is likely to be the result of increased productivity (through the discussed spillovers), and through an increased demand for labor. Because foreign plants also have higher productivity and pay higher wages than local firms, the two factors imply that a higher presence of foreign firms raises the general productivity and wage level in a province or region.
[46] Modernization theorists, for example, can argue that a rising middle class (due in large part to the regime's successful development policies) demanded political liberalization and a transition to democracy. Economists can claim that policy mismanagement in the midst of economic collapse accelerated the regime's downfall, while political economists (with a Marxist perspective) can attribute the regime's demise to the state's dependence on capital and mobile capital flight. Historians interested in the influences of great men can look to Suharto's failing health (he suffered two strokes in the lead up and during the Asian Financial Crisis). Finally, international relations scholars can appeal to post–Cold War politics, and specifically to changing American preferences for supporting the Suharto regime (e.g., during the economic crisis, media reports suggest that US Secretary of State Madeline Albright phoned Suharto to recommend he resign).

authoritarian regimes in Indonesia and Malaysia with the Asian Financial Crisis, Pepinsky (2008, 349) argues that the composition of regime supporters affected each government's policy response:

> Indonesia's attempt to repeg the rupiah and political conflict over capital account closure reflect contradictory preferences for adjustment policies among the regime's supporters – fixed and mobile capital – which ultimately helped to bring down the New Order in May 1998. Malaysia's implementation of capital controls and an exchange rate peg reflects the regime's dependence on an alliance between Malay fixed capital and Malay labor, which had complementary preferences over adjustment policy. This facilitated Mahathir's task of remaining in power despite Malaysia's economic crisis, for the regime chose policies that its supporters demanded.

The Suharto regime's commitment to protect its regime supporters underlay its decision – often contradictory – to float the exchange rate (in an effort to earn the respect of international investors and lending agencies), pressure the Bank of Indonesia to raise interest rates (in order to discourage capital flight and inflation), and selectively cut public projects (in order to gain financial support from the international financial community, in particular the International Monetary Fund). Interestingly, projects that were connected to the interests of Suharto's family, a close crony, or the military were not cut.[47] Moreover, throughout 1997 and 1998 regime supporters pressured the Suharto regime to adopt favorable responses to the crisis. The regime did so by extending corporate bailouts, while cutting subsidies for fuel and electricity.

By mid-1998, these policy decisions contributed to rising inflation, further currency depreciation, bank runs, and an overall deterioration of economic conditions. The economy contracted 14.8 percent from the previous year and political conflict among members of the New Order's support coalition arose. A final round of subsidy cuts led to massive riots by urban residents of Jakarta and other major cities on May 13–14, 1998. These riots led to more than 1,000 deaths.[48] A week later, on May 21, 1998, General Suharto resigned and the New Order came to an end.

6.7 CONCLUSION

Dictatorships that extract rents from economic production can enhance their prospects for political survival. This strategy is particularly prevalent in industries that require high fixed costs and capital investments. This chapter presents cross-national evidence that FDI associated with oil production – an industry with high fixed costs and capital requirements – can generate rents that an autocrat can use to finance the military. To identify a causal relationship,

---

[47] Soesastro and Basri (1998).
[48] Observers disagree about who (if anyone) instigated these riots, but possible culprits were all members of the army's highest leadership (Pepinsky 2008, 464).

6.7 Conclusion

I exploit the "randomness" of oil discoveries to evaluate how *new* opportunities for rents leads to higher inflows of FDI *and* military spending in *autocracies*. In doing so, FDI can help an autocrat retain the loyalty of an important domestic ally.

In addition to financing the military, many dictators actively court FDI as a part of a broader strategy to foster economic growth that benefits a larger swath of the population. These dual benefits of inward FDI underlay General Suharto's governing strategy in Indonesia for over three decades. From the mid-1960s through the mid-1990s, FDI generated rents for Suharto's inner circle of supporters (i.e., the military, his family) and sparked broader economic growth. However, the regime's response to the 1997 Asian Financial Crisis contributed to a precipitous drop in FDI, an outflow of foreign capital that accelerated the demise of Suharto's dictatorship. This last point highlights a potential danger of capital outflows: while capital inflows can empower autocrats through the generation of rents, a sudden capital outflow can potentially accelerate a dictatorship's demise.

7

Conclusion

Cross-border flows of foreign aid, foreign direct investment (FDI), and remittances are salient features of the global economy. These transfers involve different actors – firms, governments, and households – in almost every country. For many countries, these capital flows account for a significant share of national income that can potentially affect politics. This raises an important question: can governments harness these foreign capital flows to their political advantage? And if this is the case, how might governments do so? These are the central questions addressed by this book.

The "answer" in a nutshell is that *nondemocratic* governments can leverage international capital flows to their political advantage. These governments incorporate foreign capital into their revenue base to fund strategies that bolster their political survival. This final chapter briefly summarizes the book's arguments and evidence. It then discusses some potential extensions for future research and the welfare implications of international capital.

7.1 THE ARGUMENT AND EVIDENCE

The Argument

Chapter 3 presented the book's central theoretical argument that governments can incorporate international capital to fund strategies of political survival, with effects that are magnified in nondemocracies. In autocracies, these strategies are fostering *repression* and accumulating *loyalty* through targeted transfers to a small group of regime loyalists (patronage).

All governments derive their revenue through a combination of tax and nontax sources. And it is through the latter that a government can incorporate foreign aid, remittances, and FDI in its revenue base. In this book, I developed a "unified" model of international capital and political survival that

7.1 The Argument and Evidence

identifies three distinct channels by which foreign capital can serve as a source of nontax income. A government can siphon some fraction of foreign aid directly into its revenue base, representing an income effect. In contrast, remittance income – which is largely untaxed household income – can allow a government to reduce its provision of welfare goods and divert those expenditures to financing repression and patronage. This expenditure-switching strategy represents a substitution effect. FDI can increase a government's revenue base in two ways: through higher wages that a government can tax and by creating opportunities for government rents. The latter of these channels represents a rentier effect. Moreover, as Chapter 3 argued, the incentive for governments to utilize these channels becomes *magnified* in less democratic countries. This led to the book's central conjecture: that dictatorial rule can be magnified by international capital.

The Evidence

Because all governments must survive to achieve their objectives, a simple test of the book's central conjecture was to analyze the relationship between international capital inflows and the political tenure of governments (leaders) across nondemocratic and democratic regimes. Chapter 2 evaluated this conjecture. In a series of bivariate plots of just the raw data, I demonstrated a positive association between the duration of governments and capital inflows in countries with more authoritarian political institutions. I corroborated these patterns with more robust statistical analysis, demonstrating that capital inflows are associated with a lower likelihood of leaders losing power in countries with less democratic political institutions.

On the surface, these associations demonstrate a relationship between capital inflows and authoritarian governance. However, such associations do not necessarily establish a causal relationship between these variables nor do they identify the underlying channels associated with each type of capital inflow (e.g., a substitution effect associated with remittances). The ability to draw causal inferences, in particular, is statistically biased by the endogeneity of foreign capital with politics.

The cross-national empirical analysis in Chapters 4, 5, and 6 addressed these concerns by "identifying" the causal impact of each type of foreign capital on various dimensions of authoritarian politics. Chapters 4 and 5 leveraged instrumental variable research designs to show that US foreign aid harms political rights by lowering a government's tax effort (Chapter 4), while remittances expand the political powers of dictators by allowing the leader to divert expenditures from welfare goods to patronage (Chapter 5). Chapter 6 employed a different identification strategy, by gauging the impact of inward FDI on military spending using the randomness of oil discoveries. In autocracies, FDI causes a government to increase its expenditures on the military, while there is no effect in democracies. Table 7.1 summarizes the

TABLE 7.1 *The perils of international capital: Summary of findings*

			Outcome variables	
Capital	Channel	Exogenous variation	Autocracy	Government finance (channel)
(1)	(2)	(3)	(4)	(5)
US foreign aid	Income effect	Legislative fragmentation × Probability a country receives foreign aid	Political rights	Tax effort (−)
			Executive constraints	Transfers (+)
			Leader exit	Military spending (−)
Remittances	Substitution effect	World oil price × Muslim recipient country distance to Mecca	Executive constraints	Corruption (+)
			POLITY	Government transfers (−)
			Leader exit	Public sector compensation (+)
Foreign direct investment	Rentier effect	"Random" oil discoveries	Leader exit	Military spending (+)
			Coup	
			Civil war	

Notes: In column 5, government transfers refer to government expenditures on welfare payments (e.g., subsidies and transfers). In column 5, "+" indicates whether the associated type of capital (in column 1) is positively related to the proposed channel (in column 5). In contrast, "−" indicates whether the associated type of capital is negatively related to the proposed channel.

book's empirical strategy and evidence. It lists each capital flow, its associated source of exogenous variation, and the outcome variables used to measure the causal impact of that capital flow on autocracy and the proposed channel.

The statistical analysis in each chapter was corroborated by country vignettes describing how each type of foreign capital can affect authoritarian politics. The analysis of General Siad Barre's governing strategy in Somalia in Chapter 4, for example, shows how an autocrat can use foreign aid – particularly from the United States – to fund his repressive dictatorial rule. In Chapter 5, I analyzed the impact of remittances on authoritarian politics in Egypt and Jordan. This comparative vignette revealed that during the 2011 Arab Spring, sufficiently buoyant remittance income permitted Jordan's King Hussein to withstand the political upheaval that spread across the Arab region. By contrast, declining remittance flows to Egypt undermined the Mubarak regime's capacity to thwart its ouster from power.

7.1 The Argument and Evidence

Analogous to remittance income, a sudden drop in FDI can also serve to delegitimize an autocratic regime and expedite its demise. As Chapter 6 describes, General Suharto actively courted FDI as means to sustain his dictatorial rule in Indonesia. Foreign investments, especially in the country's commodity sectors, spurred economic growth and provided opportunities for patronage and corruption. When this capital fled – as a consequence of the Asian Financial Crisis – it accelerated the government's fall from power.

Drawing the Appropriate Inferences

The evidence marshaled in Chapters 4 to 6 painted a narrative that international capital can sustain and expand authoritarian politics. Of course, as with all social science research, one should be cognizant of drawing the appropriate inferences. With this in mind, each chapter carefully provided the scope conditions for interpreting the causal effects. For example, Chapter 4 showed that US aid causes repression, primarily in autocracies. And while the United States is the world's largest bilateral aid donor, similar causal inferences may not necessarily apply to aid from other donors. Future research may, therefore, want to identify the causal effects of aid from other donors on repression, possibly by leveraging similar measures of political fragmentation on the aid allocation process as the basis of instrumental variables for these donors.[1]

Whereas Chapter 4 gauged the causal impact of US aid on autocratic governance in about 130 countries, Chapters 5 and 6 facilitated causal inferences involving a smaller subset of countries. In particular, Chapter 5 leveraged a quasi-natural experiment to identify the causal impact of oil price–driven remittance inflows on autocratic governance in non–oil-producing Muslim-majority countries, while Chapter 6 exploited the randomness of oil discoveries to evaluate the causal effect of FDI on military spending in countries producing oil. The limited set of recipient countries in each chapter poses challenges to generalizing the causal inferences to a broader set of countries. Accordingly, each chapter transparently qualified limitations (applicability) of its inferences and the potential for generalizability beyond the treatment group of countries that were utilized (e.g., oil producers in Chapter 6).

For example, Chapter 5 demonstrated that Muslim-majority non–oil producers were "comparable" to similarly situated non–oil-producing non-Muslim countries on various observable economic and political characteristics

---

[1] Building on the identification strategy in Ahmed (2016), scholars have used alternate measures of legislative fragmentation to construct instrumental variables for additional donors. Dreher and Langlotz (2017), for example, measure donor fractionalization as the probability that two randomly chosen deputies from different parties in the same governing coalition to instrument for foreign aid. This instrument is apt to identify government fractionalization in parliamentary democracies. Using this instrument, Dreher and Langlotz (2017) evaluate the impact of aid on growth, while Langlotz and Potrafke (2017) analyze whether aid affects military spending.

(e.g., GDP per capita, POLITY2 score, conflict) *prior* to the treatment (i.e., oil price–induced remittance inflows). As such, it is plausible to infer that *had* non-Muslim non–oil producers experienced a similar "shock" to remittance income, these countries would also have become more authoritarian (i.e., experienced reductions in executive constraints). However, drawing more stringent causal inferences beyond this would necessitate identifying additional and possibly broader-ranging sources of exogenous variation in international capital flows.

Of course, identifying such exogeneity is not easy, but it does represent a worthy endeavor for scholars to pursue in order to better understand the causal impact of international capital (and, more broadly, of globalization) on politics. This book serves to promote such scholarship by presenting three distinct research designs to evaluate the causal impact of foreign aid, remittances, and FDI on politics for large samples of countries around the world. Moreover, I encourage future research to further evaluate the arguments in this book with innovative and carefully executed identification strategies.

7.2 EXTENSIONS

The arguments in this book can be extended and assessed in various ways beyond the performance of further cross-national analyses evaluating the causal impact of international capital on politics. Some of these possibilities are further described in the text that follows.

Subnational Analysis

One possible extension could evaluate whether international capital affects politics at the subnational level. For example, do mayors and governors lower their provision of local public services in cities (or provinces) with a higher share of remittance-receiving households? With respect to FDI, the location of these projects is often not homogenously distributed within recipient countries, with investments instead being typically concentrated in urban centers and near ports. Accordingly, a key question is whether FDI can generate rents that accrue to local leaders and government officials in regions with high amounts of FDI inflows. And while foreign aid is generally thought to represent a transfer between national governments, in practice the allocation of funds by a recipient government is often geographically targeted within their countries, with the objective to preserve the incumbent's rule. For example, an incumbent in a "hybrid" autocracy with elections could use foreign aid to finance preelectoral repression, buy electoral support (e.g., vote-buy), or reward districts that vote for the incumbent with "pork projects." Examining the subnational allocation of aid thus offers an opportunity to evaluate how an incumbent government can preserve its grip on power.

7.2 Extensions

Motivated by these types of questions, scholars in the past few years have been increasingly collecting subnational data – which is often quite challenging in relatively "closed-off" autocratic regimes – to better understand subnational governance.[2] And several recent studies have investigated the impact of international capital on subnational politics in autocracies, with findings that are largely consistent with the cross-national findings presented in this book.[3] For example, Zhu (2017) asks whether activities by multinational corporations (MNCs) in Chinese provinces generate opportunities for rent extraction in the form of corruption (e.g., bribery) among subnational government officials, such as senior cadres of the Communist Party.[4] Using various measures of corruption (e.g., number of filed cases, value of recovered funds, "perceptions" by individuals and firms), Zhu provides strong statistical evidence that provinces with more MNC activity are strongly associated with higher levels of corruption and that firms in regions with greater MNC activity tend to pay more bribes. Moreover, with an autocratic central government, provincial leaders may leverage foreign investments in such manners that advance their political careers and that possibly allow them to exercise greater autonomy from the central government.[5]

---

[2] Many of these research designs attempt to evaluate causal effects and include the use of instrumental variables, regression-discontinuity designs, and differences-in-differences (DID). Zhu (2017), for example, uses instrumental variables to gauge the impact of MNC activity on provincial-level corruption. Jablonski (2014) uses DID to evaluate how a change in government affects the geographic allocation and level of foreign aid in Kenya.

[3] FDI can also have differential and perverse consequences at the subnational level in democracies. Rudra et al. (2018), for example, study the distributional effects of FDI in water-intensive sectors in India. They identify "conditional" effects at the subnational level. Areas marked by relatively large poor and marginalized populations – where regulatory capture is more likely – fare worse. Individuals in these areas experience a slower rate of access to potable water (a resource that is critical to day-to-day welfare).

[4] Such an evaluation is challenging, as it is notoriously difficult to identify and objectively measure corruption (Treisman 2007). Zhu addresses this challenge in multiple ways. He relies on "objective" corruption cases reported by the procuratorate (*jian cha yuan*) and collects data on the number of corruption cases filed, the amount of recovered funds, and the number of senior cadres disciplined (at or above the county or division level) for each of China's provinces in 1998 and 2002. Moreover, for robustness Zhu uses survey data from both individuals and firms as additional measures of corruption. For the former, Zhu uses the frequency that residents witnessed corruption and the level of perceived corruption. On the latter, Zhu exploits a large firm survey conducted by China's National Bureau of Statistics and uses firms' expenditures on entertainment and travel costs as a proxy for bribes. The analysis of these data is consistent with the main findings in Chapter 6.

[5] For example, to evaluate the latter in an autocracy, Malesky (2008) analyzes news stories from six daily newspapers between 1990 and 2000 to measure a central government's disapproval of behavior by local leaders across Vietnam's sixty-one provinces. Using a simultaneous equation estimation approach to address concerns with endogeneity, Malesky finds that greater amounts of FDI in a province empowers local leaders to willfully push beyond central economic or administrative policy ("red lights"). In doing so, Malesky concludes that "FDI appears to have a powerful and robust impact on de facto decentralization" (115), thus empowering subnational leaders in Vietnam.

Other forms of international capital can also affect subnational politics. For instance, project-level foreign aid data have become available in the past decade, allowing researchers to investigate the strategic distribution of aid by leaders and the ensuing political consequences.[6] For example, in a recent study, Jablonski (2014) exploited data on the geographic and temporal distribution of World Bank and African Development projects in Kenya from 1992 to 2010, a period during which the country successfully transitioned from autocratic to nascent democratic governance.[7] Using a differences-in-differences research design, Jablonski demonstrates that with new regime changes in Kenya, the geographic distribution of aid spending shifts toward the new regime's support base and frequently away from those of the losing regime.[8] These findings support a political model of aid allocation in which electoral strategies play a strong and consistent role in aid spending by an incumbent. Such a model is consistent with the income effect advanced in this book.

Whereas some scholarship evaluates the impact of FDI and foreign aid on subnational governance in an autocratic regime (e.g., China, Vietnam) or a nascent democratic one (e.g., Kenya), similar studies for remittances are scant. In large part this reflects the difficulty in collecting data on remittances (e.g., monetary value, how it is spent) at the subnational level in an autocracy, where such data may not be accurately collected or available for analysis by researchers.[9] That said, scholars may be able to triangulate different data sources to better understand how remittances can affect individual preferences/attitudes and the behavior of local governments (e.g., tax effort, provision of public services). For instance, in ongoing research that builds on the findings in Chapter 5, I examine whether municipal governments in the Kingdom of Jordan follow a substitution effect logic in response to remittance income.[10]

In particular, I map a nationally representative labor survey (that asks individuals whether they receive any remittances and if so, their monetary value) to municipal-level data on public spending, demographics, socioeconomic, and political characteristics. A key finding is that municipal governments collect fewer taxes – a common measure of local government effort in the provision of public services in developing countries – in areas where a higher share of remittance-receiving

---

[6] For example, see Dreher et al. (2016); Jablonski (2014).
[7] During the 1990s, Kenya transitioned from an "intermediate autocracy" (with a POLITY2 score of −5 in 1992) to a "weak autocracy" in the early 2000s (with a POLITY2 score of −2 in 2001). By 2002, Kenya had fully transitioned to democracy with a POLITY2 score of +8.
[8] Moreover, Jablonski provides evidence that geographic allocation of aid affects electoral outcomes, frequently helping the incumbent win elections.
[9] There is a burgeoning literature evaluating how local governments in *democracies* – most notably Mexico since 2000 – "respond" to remittances. One strand of this research investigates how local governments augment their spending in relation to Mexico's 3×1 program (Simpser et al. 2016).
[10] Ahmed (2017b).

households.[11] This finding is largely consistent with a substitution effect: additional household income in the form of remittances permits a local government to reduce its provision of certain public goods (in this case, tax collection) with the expectation the household can plausibly finance those forgone public services for itself.

The foregoing studies provide preliminary evidence that FDI, foreign aid, and remittances can affect subnational politics in autocracies (and nascent democracies) in ways that are broadly consistent with the theoretical channels developed in Chapter 3. This is reassuring; however, additional analysis at the subnational level should be pursued.[12]

Politics in Democracies

Another direction for future research is to investigate how international capital flows affect politics in hybrid and democratic regimes. Since the 1970s, the number of democracies has increased over time, and while dictators still rule over many parts of the world, they have often adopted hybrid democratic practices (e.g., legislative elections, decentralized governance at the local levels) to support their tenure in power.[13] This prompts several questions that could be probed by future research. For instance, as dictators liberalize their regimes – often reluctantly and slowly – how might the level and/or volatility of international capital accelerate or impede this process? How might these effects vary by the type, timing, and intensity of each capital inflow? How and why might these effects vary across countries with different types of democratic institutions (e.g., parliaments vs. presidential systems), by interactions across constituent groups (e.g., the military, the clergy, ethnic groups), and by varying levels of democratic consolidation?

One important dimension of democratic politics – which also features in some autocracies – is electoral competition. In autocracies with legislative elections, such as Jordan, how might international capital affect the strategic behavior of a government and opposition party candidates? And how might these effects vary by the timing, intensity, and recipient of each type of capital

---

[11] Moreover, this finding is magnified on the intensive margin: municipal governments exert less tax effort in areas where the average value of remittance income is higher. Ahmed's empirical analysis controls for various factors that can influence the municipal public finances including population size, the unemployment rate, poverty rate, transfers from the central government, and measures of local political competition (e.g., the number of tribes that competed in the last municipal election).

[12] Despite the challenges in collecting subnational data on international capital flows *and* political outcomes (e.g., survival of local leaders, public spending), especially in authoritarian regimes (where such data may not be collected or available to researchers), research in this direction is nonetheless a worthy pursuit. Moreover, in many instances, "going subnational" offers researchers an opportunity to utilize a wider array of research designs to estimate causal effects (e.g., regression discontinuity designs, differences-in-differences, instrumental variables, simultaneous equations).

[13] Ghandi (2008).

flow? For example, a donor may want to assist an incumbent and his or her affiliated party candidates' electoral prospects by increasing foreign aid disbursements in the lead-up to an election, while "politically motivated" remittances might finance the political campaigns of anti-incumbent party candidates.[14]

In addition to influencing the electoral strategies of political parties, foreign capital may also shape the preferences, attitudes, and behavior of individuals in their interaction with political actors. Scholars are increasingly probing this line of inquiry, often with novel experimental approaches to tease out causal relationships. For instance, do citizens view programs funded by foreign aid and the government *differently* from elites? And why might this be the case? Findley et al. (2017) examine this question in Uganda. They field various survey experiments and randomize the identity of donors to show that members of parliament support government programs over foreign aid, whereas citizens prefer aid over government. This finding is consistent with the conjecture that elites are better able to capture revenues from government programs (i.e., an income effect). In contrast, citizens prefer aid projects, as they may be less prone to elite capture and are accordingly more likely to generate benefits for the masses. Future research could evaluate whether this finding holds in other countries, with different local political economies, and in relation to other foreign capital flows.[15]

Other Aspects of Globalization

Another direction for future research is to integrate other aspects of globalization. This would broaden the book's central question by attempting to understand the following: how can autocrats "manage" the various facets of globalization (beyond just financial globalization in the form of FDI, foreign aid, and remittance inflows) to their political benefit? In pursuit of this broader endeavor, there are several ways future research could proceed.

---

[14] On political aid cycles, see Faye and Niehaus (2012). On political remittance cycles, see O'Mahony (2013).

[15] On the latter, research to date focuses on "attitudes" toward foreign aid, remittances, and FDI separately. For example, Ahmed (2017a) finds that remittances "mediate" how individuals view unpopular governments in Latin America. Ahmed finds that remittance recipients tend to a hold a more positive assessment of the national economy and attribute this favorably to the incumbent government. As a consequence, Ahmed finds that remittance recipients are more likely to support a disliked incumbent than a nonremittance recipient. Research on FDI tends to examine individual attitudes in favor or against foreign investment. The emerging research suggests that individuals with greater skill, education, and income hold more favorable attitude toward FDI (Pandya 2010; Ahmed et al. 2017). None of these studies examine individual attitudes toward aid, remittances, and FDI in conjunction.

7.2 Extensions

One extension would be to account for other forms of foreign capital, such as sovereign bonds and foreign portfolio investments. Such an extension is plausible without onerous adjustments to the underlying theory. For example, sovereign bonds comprise government-issued debt that generates revenue for the central government.[16] As such, sovereign bonds are akin to foreign aid receipts and in theory can enhance an autocrat's prospect of political survival through an income effect (i.e., some fraction of the revenue from sovereign bonds finances patronage and repression). In contrast, while portfolio investments are distinct from FDI, they may also augment an autocrat's revenue base.[17] The infusion of foreign portfolio investments (e.g., stocks, corporate bonds) in a country's banking system and equity markets can help domestic firms raise their productivity (and hence lead to higher profit and labor income, which in turn can be taxed by the government), as well as offer opportunities for rent extraction (e.g., if the investments are in sectors with high fixed costs). In theory, therefore, portfolio investments may generate similar revenue-enhancing opportunities for governments as those afforded by FDI. Of course, future research could empirically evaluate whether sovereign bonds and portfolio investments can also empower autocratic rule either by augmenting a government's public finances or through different channels.[18]

A second extension would be to incorporate international trade – the flow of goods and (increasingly) services across national borders – in the book's theory and empirical analysis.[19] This is a natural extension, as there is a voluminous literature on the political economy of trade. One strand of this research probes how governments forge their trade policies, while another examines how international trade affects government policy.[20] Research focused on the latter, for example, has identified relationships between trade and the size of the public sector, corruption, and even interstate conflict.[21] This second strand is particularly pertinent in understanding how international capital and trade can affect governance in recipient countries, particularly in nondemocracies.

---

[16] Of course, this debt issue comes with the expectation of repaying foreign lenders the value of the coupon plus an interest payment.

[17] A foreign portfolio investment is basically the entry of financial funds into a country where foreigners deposit money in a country's bank or make purchases within the country's stock and/or bond markets. These investments tend to be short term and often for speculative purposes.

[18] Recent research suggests that this may be the case. For instance, DiGiuseppe and Shea (2016) find that sovereign borrowing can extend the duration of leaders in nondemocracies.

[19] In the context of understanding how foreign aid, remittances, and FDI affect "politics," such an extension seems natural. On the "demand side," income from aid, remittances, and foreign investment is used by firms, governments, and households to buy goods (and services) from abroad. On the "supply side," FDI can help firms export their goods abroad (for example, by improving the quality of their products and/or lowering costs). Relatedly, immigration can also raise the productivity of firms and/or lower the price of exported goods.

[20] On the former, see Grossman and Helpman (1994) and Rodrik (1995).

[21] On trade and the public sector, see Rodrik (1998). On corruption, see Ades and Di Tella (1999). On interstate conflict, see Gartzke (2007).

An autocrat may leverage international trade in a number of ways to help him survive in office.[22] One prominent strategy is to utilize exports to stimulate economic growth and cultivate performance legitimacy. Some policies include subsidies to export competing industries, "interventions" to artificially undervalue the country's exchange rate (and thus keep exports competitively priced in foreign markets), and restrictions pertaining to imports that compete with domestic industries via tariffs and nontariff barriers (e.g., domestic content requirements). Indeed, many of these policies underlay the tremendous export-led growth of the Asian Tigers under dictatorial rule (e.g., South Korea, Indonesia) and most recently, China.

In addition to promoting political survival via economic growth, international trade could also generate revenue for a government.[23] In many autocracies, for example, state-owned enterprises (SOEs) generate profits for the government by engaging in international commerce. For instance, some of these SOEs include nationally owned oil companies in the Persian Gulf Monarchies that derive a substantial portion of their revenues through petroleum exports, as well as state-owned banks (e.g., China Investment Corporation) that reap significant financial returns by lending and investing abroad.[24] Exports could also generate revenue for an autocrat in a manner consistent with this book's theory. For instance, an autocrat could tax the revenue of export-oriented firms and the (higher) wages of their workers. Moreover, export-oriented firms that operate in sectors with high fixed costs (e.g., natural resources) and/or rely on substantial government assistance (e.g., partnerships with pharmaceutical and high-tech firms) are ripe for opportunistic rent extraction. The availability of these channels – tax collection and rent extraction – thus offers pathways for governments to increase their revenues via exports. Future research could investigate these pathways further.

A third extension is to account for the greater "institutionalization" of globalization. Since 1945, the number, scope, and sophistication of multilateral organizations as well as nongovernmental entities (e.g., associations comprised of private banks and financiers) dedicated to expanding and deepening cross-border economic transactions has grown tremendously.[25] Among international organizations (IOs) composed of nation-states, some of these include the World Bank and various regional development banks that provide foreign aid to

---

[22] These policies are not exclusive to autocrats. Democratic governments may also use trade policies to extend their political tenure.
[23] Bastiaens and Rudra (forthcoming) document how trade liberalization has lowered revenues from trade taxes (e.g., customs duties, tariffs). In many developing countries, this has triggered reductions in welfare spending, with differential effects across democracies and nondemocracies.
[24] For a recent account on the political economy of China's outward direct investment see Shi (2014).
[25] See Keohane (1984) on how international organizations help facilitate economic exchange among nations in an anarchic international system.

7.2 Extensions

governments, the World Trade Organization (previously, the General Agreement on Tariffs and Trade [GATT]) to regulate and expand international trade, and various multilateral financial institutions (e.g., the International Monetary Fund, Bank of International Settlements, regional development and investment banks) and international legal treaties that facilitate international investments (e.g., bilateral and multilateral investment treaties).[26]

In many instances, these international economic institutions can influence politics in member countries. Sometimes governments join IOs as a means to import democratic institutions, while in other instances IOs attempt to regulate the behavior of member countries and to improve their governance.[27] That said, many autocrats may be able to leverage these institutions to both increase international economic exchange and enhance their own prospects for political survival. Consider, for example, bilateral investment treaties (BITs), which have proliferated in number and scope in the past fifty years.[28] BITs aim to increase international investment between signatory countries by improving the investment climate in host countries by specifying dispute resolution mechanisms, clarifying compensation for expropriation, and raising investment entrance and profitability.

BITs may be particularly effective for developing countries – especially those with autocratic regimes – in strengthening the protection of domestic property rights as a means to attract foreign investment. Rosendorff and Shin (2011) posit that authoritarian regimes can attract FDI with BITs by "importing" the BIT's governance institutions, which in turn may extend the regime's political tenure.[29] Arias et al. (2017) provide some empirical corroboration. Using survival models, these scholars find that BITs enhance the political survival of autocrats relative to democrats by improving the domestic investment climate in autocracies.

---

[26] Poulsen (2015) provides a solid account on the international legal innovations, history, and consequences associated with bilateral investment treaties, especially in developing countries.

[27] See Pevehouse (2005) on how states can "import" democracy by joining IOs. Lodewijk and Knack (2015) examine how IOs can help improve governance in member states. For example, in an effort to increase the "effectiveness" of foreign aid in raising living standards, they document how the World Bank frequently targets projects in countries with lower levels of corruption as well as projects explicitly designed to help those governments reform their public sectors. In other settings, an IO may attempt to regulate the behavior of actors in the "sending" country. The Organisation for Economic Co-operation and Development (OECD)'s Anti-Bribery Convention, for example, aims to reduce corruption in developing countries by encouraging sanctions against bribery in international business transactions conducted by companies *based in the Convention member countries*. The convention criminalizes acts of offering or paying bribes.

[28] Elkins et al. (2006); Busch and Tobin (2010); Rose-Ackerman and Tobin (2011).

[29] Recent research suggests that this effect may be mediated by the quality of domestic governance (institutions) within an autocracy. Surveying a sample of US-based multinational corporations and analyzing a cross-national panel of countries, Bastiaens (2016) finds BITs increase FDI in autocracies that allow some degree of "public deliberation" in the formation of economic policies. For international investors, public deliberation signals greater transparency and commitment to property rights enforcement. Autocracies with less (or no) public deliberation attract less FDI.

BITs, of course, are just one type of international economic institution that autocrats can potentially leverage to their political advantage. Autocrats may also be able to exploit IOs dealing with aid, trade, migration, and other aspects of globalization to their benefit. This is a fruitful area of future research. More broadly, the three extensions outlined earlier should not be viewed in isolation. They are likely to affect politics in combination with each other, as well as with the three types of international capital examined at the core of this book. Importantly, these extensions should not be viewed as being exhaustive of the channels through which international capital can affect politics in an increasingly dynamic and interconnected global political economy.[30]

7.3 IS INTERNATIONAL CAPITAL BENEFICIAL?

The potential theoretical and empirical extensions to the central arguments and insights in this book are encouraging and exciting! That said, the book's welfare implications are *potentially* quite grim. Indeed, a casual reading of the book's central argument implies that international capital is politically perilous in *all* contexts. This, however, misses the conditional nature of the book's central argument: a country's political institutions *mediate* the effect of foreign aid, remittances, and FDI on politics. In particular, as elaborated in Chapter 3, the *institutional features* in nondemocratic regimes provides the *incentives* for governments to spend the rents from international capital on repression and on targeted benefits to regime loyalists. In nondemocracies, these two strategies are effective in sustaining authoritarian politics.

The conditional effect of international capital across regime types implies that if a recipient country's institutions *change*, so can the impact of international capital flows on politics. While political institutions are "sticky," and can be made stickier with foreign capital (as this book demonstrates), they are not immutable. A country's underlying political institutions *can* change, due to both "endogenous" (e.g., economic development, inequality) and "exogenous" (e.g., foreign "interventions") factors.[31] And, *if* these factors can push a country to adopt more democratic institutions, then the pernicious effects of international capital on politics can wane.

With the rise in the number of democracies worldwide since 1800, especially over the sample period examined in this book (1970–2015), such

---

[30] Some possible extensions include the role of international security (e.g., terrorism, civil conflict), changing geopolitics (e.g., "rise" of China and India), geography, and the climate.

[31] Some examples of endogenous factors include the level of economic development, inequality, and associated effects of internal civil unrest (e.g., Boix and Stokes 2003; Acemoglu and Robinson 2006). Some examples of exogenous factors include changes in the structure of the international system (e.g., the distribution of power among the Great Powers), "interventions" by external powers, and the regional "diffusion" of governance institutions and norms (e.g., Huntington 1991; Aidt and Albornoz 2011; Boix 2011).

7.3 Is International Capital Beneficial?

a perspective is encouraging. This perspective features prominently in how many policymakers and academics think about international capital in economic and political development. Take, for example, some contemporary debates on the "effectiveness" of foreign aid in development. An increasingly popular notion is that aid fosters economic development in countries with democratic institutions – a perspective that had led donors to reward democratic recipients and those committed to political liberalization with greater amounts of aid.[32]

Whether foreign aid actually causes growth in democratic recipients remains unclear.[33] Nevertheless, the lively policy discussions about "aid effectiveness" relate to a broader theme in this book: the effect of international capital on political and economic welfare.[34] While both are inherently intertwined, this book has primarily focused on the latter. On that front, the theory and empirics show that international capital received in autocracies is politically perilous. What about the economic welfare consequences associated with international capital in autocracies? Answering this question is complicated, as there are both direct and indirect effects.

On the former, remittances and FDI directly raise individual (household) income thus permitting individuals to consume more of the goods they want.[35] This unambiguously raises individual economic welfare. The *indirect* economic welfare effects associated with international capital are less clear, as they stem primarily from the consequences of government policy. On the one hand, some fraction of the rents captured by a government from capital inflows can be spent on welfare goods (e.g., health care, education) that benefit the masses. On the other hand, the fraction of government revenue spent on welfare goods is lower in autocracies. Thus, economic welfare for the masses (via welfare goods) is lower in autocracies than in democracies. Of course, whatever the government does not spend on welfare goods can be spent on targeted transfers to a smaller group of individuals (e.g., military leaders, coethnics, party loyalists). The recipients of such patronage benefit economically, and more so in less democratic countries. As the quality of democratic institutions varies across countries (and time), so do the indirect

---

[32] Burnside and Dollar (2000); Bermeo (2011). Bermeo finds donor "characteristics" shape the allocation and potential effects of aid on governance. In particular, aid from democratic donors is positively associated with democratic governance in recipients, while aid from autocratic donors is negatively associated with democratic governance. Bermeo does *not*, however, establish a causal relationship between aid and governance.

[33] For example, Dreher and Langlotz (2017) leverage an instrumental variable research design to show that foreign aid does not cause growth across a large sample of countries, including a sample restricted only to democracies.

[34] Welfare can be evaluated at the individual and societal level.

[35] Remittances are a transfer of income from abroad – typically sent from a family member in another country – that directly enters a household's budget. In contrast, FDI raises the output of firms and productivity of workers and this higher labor productivity can translate to higher wages. The formal model in the appendix to Chapter 3 explicitly shows these direct effects.

economic benefits for the masses and regime loyalists. In short, while quantification of the net indirect economic welfare effects is muddled, the indirect economic welfare benefits for the masses will be lower in autocracies.[36]

Another dimension to consider involves the short-versus long-run welfare consequences. While additional income from remittances and FDI (via higher wages) can raise individual utility (e.g., greater consumption) in the short run, the entrenchment of authoritarian politics may be more deleterious in the long run. This can manifest through the underprovision of welfare goods and lower investments on public infrastructure. These welfare comparisons are further complicated by how individuals compare economic welfare in relation to political welfare. It is quite plausible that poor individuals care mainly about their economic welfare, while richer individuals place greater weight on political welfare (having satisfied their economic needs); accordingly, comparisons between economic and political welfare will vary across individual and country income.[37] These are interesting questions that are ripe for future scholarship on the economic consequences of financial globalization. With the regard to the *political* ramifications of financial globalization, as this book demonstrates, international capital can put nondemocracies in peril.

---

[36] This is consistent with the comparative statics given by Eq. (3.A11) in the appendix to Chapter 3.
[37] This is consistent with the notion that countries with higher per capita GDP are democratic and/or more likely to undergo democratization (Boix and Stokes 2003).

References

Abdih, Yasser, Ralph Chami, Jihad Dagher, and Peter Montiel. 2012. "Remittances and Institutions: Are Remittances a Curse?" *World Development*, 40(4): 657–666.

Acemoglu, Daron, and Simon Johnson. 2005. "Unbundling Institutions." *Journal of Political Economy*, 113(5): 949–995.

Acemoglu, Daron, Simon Johnson, and James A. Robinson. 2001. "The Colonial Origins of Comparative Development." *American Economic Review*, 91(5): 1369–1401.

Acemoglu, Daron, and James A. Robinson. 2006. *The Economic Origins of Dictatorship and Democracy*. New York: Cambridge University Press.

Acosta, Pablo, Pablo Fajnzylber, and J. Humberto Lopez. 2007. "The Impact of Remittances on Poverty and Human Capital: Evidence from Latin America." World Bank Policy Research Working Paper 4247. Washington, DC: World Bank.

Adam, Hussein M. 1999. "Somali Civil Wars." In Taisier M. Ali and Robert O. Matthews (eds.), *Civil Wars in Africa: Roots and Resolution*. Montreal: McGill-Queen's University Press, 167–192.

Ades, Alberto, and Rafael Di Tella. 1999. "Rents, Competition and Corruption." *American Economic Review*, 89(4): 982–994.

Adida, Claire L., and Desha Girod. 2011. "Do Migrants Improve Their Hometowns? Remittances and Access to Public Services in Mexico, 1995–2000." *Comparative Political Studies*, 44(1): 3–27.

Ahmed, Faisal Z. 2012. "The Perils of Unearned Foreign Income: Aid, Remittances, and Government Survival." *American Political Science Review*, 106: 146–165.

2013. "Remittances Deteriorate Governance." *The Review of Economics and Statistics*, 95(4): 1166–1182.

2016. "Does Foreign Aid Harm Political Rights? Evidence from U.S. Aid." *Quarterly Journal of Political Science*, 11(2): 118–217.

2017a. "Remittances and Incumbency: Theory and Evidence." *Economics and Politics*, 29(1): 22–47.

2017b. "Remitting the Democratic Deficit." Working Paper, Princeton University.

Ahmed, Faisal Z., Daniel Schwab, and Eric D. Werker. 2018. "Foreign Transfers and Comparative Democratization and Conflict." Working Paper, Princeton University.

Ahmed, Faisal Z., and Eric D. Werker. 2015. "Aid and the Rise and Fall of Conflict in the Muslim World." *Quarterly Journal of Political Science* 10(2): 155–186.

Aidt, Toke S., and Facundo Albornoz. 2011. "Political Regimes and Foreign intervention." *Journal of Development Economics*, 94(2): 192–201.

Alesina, Alberto, and David Dollar. 2000. "Who Gives Aid to Whom and Why?" *Journal of Economic Growth*, 5(1): 33–63.

Alesina, Alberto, and Roberto Perotti. 1996. "Fiscal Discipline and the Budget Process." *American Economic Review*, Papers and Proceedings, 401–407.

Alesina, Alberto, and Howard Rosenthal. 1995. *Partisan Politics, Divided Government and the Economy*. New York: Cambridge University Press.

Alesina, Alberto, and Guido Tabellini. 1990. "A Positive Theory of Fiscal Deficits and Government Debt." *Review of Economic Studies*, 57(July): 403–414.

Alesina, Alberto, and Beatrice Weder. 2002. "Do Corrupt Governments Receive Less Foreign Aid?" *American Economic Review*, 92(4): 1126–1137.

Ahlquist, John. 2006. "Economic Policy, Institutions, and Capitals: Portfolio and Direct Investment Flows in Developing Countries." *International Studies Quarterly*, 50: 681–704.

Anderson, Lisa. 1999. "Empirical Strategies in Labor Economics." In Orley Arshenfelter and David Card (eds.), *Handbook of Labor Economics*. Amsterdam: Elsevier, 1277–1366.

2011. "Demystifying the Arab Spring: Parsing the Differences between Tunisia, Egypt, and Libya." *Foreign Affairs*, 90(3): 2–7.

Angrist, Joshua D., and Alan B. Krueger. 1999. "Empirical Strategies in Labor Economics." In Orley Ashenfelter and David Card (eds.), Handbook of Labor Economics. Amsterdam: Elsevier, 1277–1366.

Angrist, Joshua D., and Jorn-Steffen Pischke. 2009. *Most Harmless Econometrics: An Empiricist's Companion*. Princeton, NJ: Princeton University Press.

Apodaca, Clair. 2002. "The Globalization of Capital in East and Southeast Asia: Measuring the Impact on Human Rights Standards." *Asian Survey*, 42(6): 883–905.

Arafat, Alaa Al-Din. 2009. *The Mubarak Leadership and the Future of Democracy in Egypt*. New York: Palgrave and Macmillan.

Arias, Eric, James R. Hollyer, and B. Peter Rosendorff. 2017. "Cooperative Autocracies: Leader Survival, Creditworthiness, and Bilateral Investment Treaties." Working Paper, New York University.

Asiedu, Elizabeth, and Donald Lien. 2011. "Democracy, Foreign Direct Investment and Natural Resources." *Journal of International Economics*, 84(1): 99–111.

Bain, Joe Staten. 1956. *Barriers to New Competition, Their Character and Consequences in Manufacturing Industries*. Cambridge, MA: Harvard University Press.

Baldwin, David. 1986. *Economic Statecraft*. Princeton, NJ: Princeton University Press.

Banks, Arthur S. 2010. "Cross-National Time Series Data Archive." Databanks International, Jerusalem, Israel.

Bardhan, Pranab. 1997. "Corruption and Development: A Review of Issues." *Journal of Economic Literature*, 35(3): 1320–1346.

Bastiaens, Ida. 2016. "The Politics of Foreign Direct Investment in Authoritarian Regimes." *International Interactions*, 42(2): 140–171.

References

Bastiaens, Ida, and Nita Rudra. 2018. *Democracies in Peril: Taxation and Redistribution in the Global Economy*. Cambridge: Cambridge University Press.

Bearce, David H., and Seungbin Park. 2017. "Why Remittances Are a Political Blessing and Not a Curse." Working Paper, University of Colorado Boulder.

Beblawi, Hazem. 1987. "The Rentier State in the Arab World." In Hazem Beblawi and Giacomo Luciani (eds.), *The Rentier State*. New York: Croom Helm, 49–62.

Beck, Thorsten, George Clarke, Alberto Groff, Philip Keefer, and Patrick Walsh. 2010. The Database of Political Institutions. Washington, DC: World Bank.

Bermeo, Sarah B. 2011. "Foreign Aid and Regime Change: A Role for Donor Intent." *World Development*, 39(11): 2021–2031.

  2016. "Aid Is Not Oil: Donor Utility, Heterogeneous Aid, and the Aid-Democratization Relationship." *International Organization*, 70(1): 1–32.

  2017. "Aid Allocation and Targeted Development in an Increasingly Connected World." *International Organization*, 71(Fall): 735–766.

Besley, Timothy, and Torsten Persson. 2010. "State Capacity, Conflict, and Development." *Econometrica*, 78(1): 1–34.

  2011. "The Logic of Political Violence." *Quarterly Journal of Economics*, 126: 1411–1445.

Besteman, Catherine. 1996. "Violent Politics and the Politics of Violence: The Dissolution of the Somali Nation-State." *American Ethnologist*, 23(3): 579–596.

Biglaiser, G., and K. DeRouen. 2010. "The Effects of IMF Programs on U.S. Foreign Direct Investment in the Developing World." *Review of International Organizations*, 5(1): 73–95.

Blalock, Garrick, and Paul J. Gertler. 2008. "Welfare Gains from Foreign Direct Investment through Technology Transfers to Local Suppliers." *Journal of International Economics*, 74(2): 402–421.

Blanton, Shannon Lindsey, and Robert G. Blanton. 2007. "What Attracts Foreign Investors? An Examination of Human Rights and Foreign Direct Investment." *Journal of Politics*, 69(1): 143–155.

Boix, Carles. 2003. *Democracy and Redistribution*. New York: Cambridge University Press.

  2011. "Democracy, Development, and the International System." *American Political Science Review*, 105(4): 809–828.

Boix, Carles, Michael K. Miller, and Sebastian Rosato. 2013. "A Complete Data Set of Political Regimes, 1800–2007." *Comparative Political Studies*, 46(12): 1523–1554.

Boix, Carles, and Susan C. Stokes. 2003. "Endogenous Democratization." *World Politics*, 55 (July): 517–549.

Boschini, Anne, and Anders Olofsgard. 2007. "Foreign Aid: An Instrument for Fighting Communism?" *Journal of Development Studies*, 43(4): 622–648.

Box-Steffensmeier, Janet M., and Bradford S. Jones. 1997. "Time Is of the Essence: Event History Models in Political Science." *American Journal of Political Science*, 41(4): 1414–1461.

Brambor, Thomas, William R. Clark, and Matt Golder. 2006. "Understanding Interaction Models: Improving Empirical Analyses." *Political Analysis*, 14(1): 63–82.

Brennan, Geoffrey, and James M. Buchanan. 1980. *Foundations of a Fiscal Constitution*, New York: Cambridge University Press.

Broadway, Robin. 2008. "Intergenerational Redistributive Transfers: Efficiency and Equity." In E. Ahmad and G. Brosio (eds.), *Handbook of Fiscal Federalism*. Northampton, MA: Edward Elgar, 355–380.

Brynen, Rex. 1992. "Economic Crisis and Post-rentier Democratization in the Arab World: The Case of Jordan." *Canadian Journal of Political Science*, 1: 69–97.

Bueno de Mesquita, Bruce, Randolph Siverson, Alastair Smith, and James Morrow. 2003. *The Logic of Political Survival*. Boston: MIT Press.

Bueno de Mesquita, Bruce, and Alastair Smith. 2009. "A Political Economy of Foreign Aid." *International Organization*, 63(2): 309–340.

2010a. "Leader Survival, Revolutions, and the Nature of Government Finance." *American Journal of Political Science*, 54(4): 936–950.

2010b. "The Pernicious Consequence of UN Security Council Membership." *Journal of Conflict Resolution*, 54(5): 667–686.

Burnside, Craig, and David Dollar. 2000. "Aid, Policies, and Growth." *American Economic Review*, 90(4): 847–868.

Busch, Mark, and Jennifer Tobin. 2010. "A BIT Is Better than a Lot: Bilateral Investment Treaties and Preferential Trade Agreements." *World Politics*, 62(1): 1–42.

Buthe, Tim, and Helen V. Milner. 2008. "The Politics of Foreign Direct Investment into Developing Countries: Increasing FDI through International Trade Agreements." *American Journal of Political Science*, 52(4): 741–776.

Campbell, James E., and Joe A. Sumners. 1990. "Presidential Coattails in Senate Elections." *American Political Science Review*, 84(2): 513–524.

Card, David. 1990. "Unexpected Inflation, Real Wages, and Employment Determination in Union Contracts." *American Economic Review*, 80(4): 669–688.

Carter, David B., and Curt Signorino. 2010. "Back to the Future: Modeling Time Dependence in Binary Data." *Political Analysis*, 18(3): 271–292.

Chami, Ralph, Adolfo Barajas, Thomas Cosimano, Connel Fullenkamp, Michael Gapen, and Peter Montiel. 2008. "The Macroeconomic Consequences of Remittances." Washington, DC: International Monetary Fund.

Chaney, Eric. 2012. "Democratic Change in the Arab World: Past and Present." *Brookings Papers on Economic Activity*, 42(1): 363–414.

Cheibub, Jose A., Jennifer Ghandhi, and James R. Vreeland. 2010. "Democracy and Dictatorship Revisited." *Public Choice*, 143: 67–101.

Chenery, Hollis B., and Alan M. Strout. 1966. "Foreign Assistance and Economic Development." *American Economic Review*, 56(4): 679–733.

Choucri, Nazli. 1986. "The Hidden Economy: A New View of Remittances in the Arab World." *World Development*, 14(6): 697–712.

Cingranelli, David, and David L. Richards. 1999. "Respect for Human Rights after the Cold War." *Journal of Peace Research*, 36(5): 511–534.

2008. "The Cingranelli-Richards (CIRI) Human Rights Data Project Coding Manual Version 2008.3.13." http://ciri.binghamton.edu/documentation/ciri_coding_guide.pdf

Clemens, Michael C., Steven Radelet, Rikhil R. Bhavnani, and Samuel Bazzi. 2012. "Counting Chickens When They Hatch: Disaggregated Aid and Growth." *The Economic Journal*, 122(561): 590–617.

Colgan, Jeff D. 2013. *Petro-Aggression: When Oil Causes War*. New York: Cambridge University.

Collier, Paul. 2010. *The Plundered Planet: Why We Must – and How We Can – Manage Nature for Global Prosperity*. New York: Oxford University Press.

Copeland, Dale C. 2014. *Economic Interdependence and War*. Princeton, NJ: Princeton University Press.

Coppedge, Michael, John Gerring, Staffan I. Lindberg, et al. 2014. "Varieties of Democracy Codebook v1." Varieties of Democracy Project: Project Documentation Paper Series. www.v-dem.net/en/

Cotet, Anca M., and Kevin K. Tsui. 2013. "Oil and Conflict: What Does the Cross Country Evidence Really Show?" *American Economic Journal – Macroeconomics*, 5(1): 49–80.

Davenport, Christian. 2007a. *State Repression and the Domestic Democratic Peace*. New York: Cambridge University Press.

 2007b. "State Repression and the Tyrannical Peace." *Journal of Peace Research*, 44 (4): 485–504.

David, A. Paul, and Gavin Wright. 1997. "Increasing Returns and the Genesis of American Resource Abundance." *Industrial and Corporate Change*, 6(2): 203–245.

de La Martinez, Jose. 2005. "Workers' Remittances to Developing Countries: A Survey of Central Banks on Selected Public Policy Issues." Policy Research Working Paper 3638, Washington, DC: World Bank.

Dietrich, Simone. 2016. "Donor Political Economies and the Pursuit of Aid Effectiveness." *International Organization*, 70(1): 65–102.

Dietrich, Simone, and Joseph Wright. 2015. "Foreign Aid Allocation Tactics and Democratic Change in Africa." *Journal of Politics*, 77(1): 216–234.

DiGiuseppe, Matthew, and Patrick E. Shea. 2015. "Sovereign Credit and the Fate of Leaders: Reassessing the Democratic Advantage." *International Studies Quarterly*, 59(3): 557–570.

 2016. "Borrow Time: Sovereign Finance, Regime Type and Leader Survival." *Economics and Politics*, 28(3): 342–367.

Djankov, Simeon, Jose G. Montalvo, and Marta Reynal-Quero. 2008. "The Curse of Aid." *Journal of Economic Growth*, 13: 169–194.

Docquier, Frederic, and Hillel Rapoport. 2006. "The Economics of Migrants' Remittances." In S. C. Kolm and J. Mercier Ythier (eds.), *Handbook of the Economics of Giving, Altruism and Reciprocity*. Handbooks in Economics (Series Editors: Kenneth Arrow and Michael Intriligator). Amsterdam: Elsevier-North Holland, 1135–1198.

Domar, Evsey. 1946. "Capital Expansion, Rate of Growth, and Employment." *Econometrica*, 14(2): 137–147.

Dreher, Axel, Andreas Fuch, Roland Hodler, Bradley C. Parks, Paul A. Raschky, and Michael J. Tierney. 2016. "Aid on Demand: African Leaders and the Geography of China's Foreign Assistance." Aiddata.org Working Paper.

Dreher, Axel, and Sarah Langlotz. 2017. "Aid and Growth: New Evidence Using an Excludable Instrument." Heidelberg University Discussion Paper 635.

Dunne, John P., and Sam Perlo-Freeman. 2003. "The Demand for Military Spending in Developing Countries." *International Review of Applied Economics*, 17(1): 23–48.

Dunning, Thad. 2004. "Conditioning the Effects of Aid: Cold War Politics, Donor Credibility, and Democracy in Africa." *International Organization*, 58(Spring): 409–423.

2008. *Crude Democracy: Natural Resource Wealth and Political Regimes*. New York: Cambridge University Press.

Easterly, William. 2006. *The White Man's Burden: Why the West's Efforts to Aid the Rest Have Done So Much Ill and So Little Good*. New York: Penguin Press.

Eichengreen, Barry, and David Leblang. 2008. "Democracy and Globalization." *Economics and Politics*, 20(3): 289–334.

Elkins, Zachary, Andrew T. Guzman, and Beth A. Simmons. 2006. "Competing for Capital: The Diffusion of Bilateral Investment Treaties, 1960–2000." *International Organization*, 60: 811–846.

Epstein, Helen. 2010. "Cruel Ethiopia." *The New York Review of Books*, May 13.

Escriba-Folch, Abel, Covadonga Meseguer, and Joseph Wright. 2015. "Remittances and Democratization." *International Studies Quarterly*, 59(3): 571–586.

Evans, Peter. 1979. *Dependent Development: The Alliance of Multinational, State, and Local Capital in Brazil*. Princeton, NJ: Princeton University Press.

Faye, Michael, and Paul Niehaus. 2012. "Political Aid Cycles." *American Economic Review*, 102(7): 3516–3530.

Findley, Michael G., Adam S. Harris, Helen V. Milner, and Daniel L. Nielson. 2017. "Elite and Mass Support for Foreign Aid versus Government Programs: Experimental Evidence from Uganda." *International Organization*, 71(4): 633-663.

Finkel, Steven E., Anibal S. Perez Linan, and Mitchell A. Seligson. 2007. "The Effects of U.S. Foreign Assistance on Democracy Building, 1990–2003." *World Politics*, 59 (3): 404–440.

Fiorina, Morris P. 1978. "Economic Retrospective Voting in American National Elections: A Micro-Analysis." *American Journal of Political Science*, 22(2): 426–443.

Fleck, Robert K., and Christopher Kilby. 2006. "How Do Political Changes Influence U.S. bilateral Aid Allocations? Evidence from Panel Data." *Review of Development Economics*, 10(2): 210–223.

Freedom House. 2011. "Freedom in the World Comparative and Historical Data." www.freedomhouse.org/report-types/freedom-world

Freeman, John R., and Dennis P. Quinn. 2012. "The Economic Origins of Democracy Reconsidered." *American Political Science Review*, 106(1): 58–80.

Frieden, Jeffry. 1991. *Debt, Development, and Democracy: Modern Political Economy and Latin America, 1965–1985*. Princeton, NJ: Princeton University Press.

Garriga, Anna C., and Brian J. Phillips. 2014. "Foreign Aid as a Signal to Investors: Protecting FDI in Post-Conflict Countries." *Journal of Conflict Resolution*, 58(2): 280–306.

Gartzke, Erik. 2007. "The Capitalist Peace." *American Journal of Political Science*, 51 (1): 166–191.

Geddes, Barbara, Joseph Wright, and Erica Frantz. 2014. "Autocratic Breakdown and Regime Transitions: A New Data Set." *Perspectives on Politics*, 12(2): 313–331.

Geroski, P. A. 1995. "What Do We Know About Entry?" *International Journal of Industrial Organization*, 13(4): 421–440.

Ghandi, Jennifer. 2008. *Political Institutions under Dictatorship*. New York: Cambridge University Press.

Ghandhi, Jennifer, and Adam Przeworski. 2006. "Cooperation, Cooptation, and Rebellion Under Dictatorships." *Economics and Politics*, 18(1): 1–26.

2007. "Authoritarian Institutions and the Survival of Autocrats." *Comparative Political Studies*, 40(11): 1279–1301.

Gibler, Douglas M., and Meredith Sarkes. 2004. "Measuring Alliances: The Correlates of War, Formal Interstate Alliance Data Set, 1816–2000" (updated through 2009). *Journal of Peace Research*, 41(2): 211–222.

Girod, Desha. 2011. "Effective Foreign Aid Following Civil War: The Nonstrategic-Desperation Hypothesis." *American Journal of Political Science*, 56(1): 188–201.

Gleditsch, Kristian S., and Michael D. Ward. 1997. "Double Take: A Re-Examination of Democracy and Autocracy in Modern Polities." *Journal of Conflict Resolution*, 41: 361–383.

Goemans, Hein, Kristian Skrede Gleditsch, and Giacomo Chiozza. 2009. "Introducing Archigos: A Dataset of Political Leaders." *Journal of Peace Research*, 46(2): 269–283.

Goodman, Gary L., and Jonathan T. Hiskey. 2008. "Exit without Leaving: Political Disengagement in High Migrant Municipalities in Mexico." *Comparative Politics*, 40(2): 169–188.

Grossman, Gene M., and Elhanan Helpman. 1994. "Protection for Sale." *American Economic Review*, 84(4): 833–850.

Guzman, Martin, Jose Antonio Ocampo, and Joseph E. Stiglitz. 2016. *Too Little, Too Late: The Quest to Resolve Sovereign Debt Crises*. New York: Columbia University Press.

Hankla, Charles R. 2013. "Fragmented Legislatures and the Budget: Analyzing Presidential Democracies." *Economics and Politics*, 25(2): 200–228.

Harrigan, Kathryn Rudie. 1981. "Barriers to Entry and Competitive Strategies." *Strategic Management Journal*, 2(4): 395–412.

Harrod, Roy F. 1939. "An Essay in Dynamic Theory." *Economic Journal*, 49(193): 14–33.

Henisz, Witold. 2000. "The Institutional Environment for Multinational Investment." *Journal of Law, Economics and Organization*, 16(2): 334–364.

Hirschman, Albert O. 1945. *National Power and the Structure of Foreign Trade*, Berkeley, CA: University of California Press.

Hochschild, Adam. 1998. *King Leopold's Ghost: A Story of Greed, Terror, and Heroism in Colonial Africa*. New York: Mariners Books.

Hoeffler, Anke, and Verity Outram. 2011. "Need, Merit or Self-Interest – What Determines the Allocation of Aid?" *Review of Development Economics*, 15(2): 237–250.

Huntington, Samuel P. 1968. *Political Order in Changing Societies*. New Haven, CT: Yale University Press.

1991. *The Third Wave: Democratization in the Late Twentieth Century*. Norman: Oklahoma University Press.

International Monetary Fund (IMF). 2012. *Direction of Trade Statistics*. Washington, DC: International Monetary Fund.

Ivanhoe, L. F. 2000. *World Oil Supply: Production, Reserves, and EOR*. Golden CO: M. King Hubert Center for Petroleum Supply Studies, Colorado School of Mines.

Jablonski, Ryan S. 2014. "How Aid Targets Votes?: The Impact of Electoral Incentives on Foreign Aid Distribution." *World Politics*, 66(2): 293–300.

Jackson, Robert, and Carl Rosberg. 1982. *Personal Rule in Black Africa*. Berkeley: University of California Press.

Jamal, Amaney A. 2012. *Of Empires and Citizens: Pro-American Democracy or No Democracy at All*. Princeton, NJ: Princeton University Press.

Jensen, Nate M. 2003. "Democratic Governance and Multinational Corporations: Political Regimes and Inflows of Foreign Direct Investment." *International Organization*, 57(3): 587–616.

2004. "Crisis, Conditions, and Capital: The Effect of International Monetary Fund Agreements on Foreign Direct Investment." *Journal of Conflict Resolution*, 48(2): 194–210.

2006. *Nation-States and the Multinational Corporation: A Political Economy of Foreign Direct Investment*. Princeton, NJ: Princeton University Press.

Jensen, Nathan M., Glen Biglaiser, Quan Li, et al. 2012. *Politics and Foreign Direct Investment*. Ann Arbor: Michigan University Press.

Jones, Benjamin F., and Benjamin A. Olken. 2009. "Hit or Miss? The Effect of Assassinations on Institutions and War." *American Economic Journal – Macroeconomics*, 1(2): 55–87.

Kaja, Ashwin, and Eric Werker. 2010. "Corporate Governance at the World Bank and the Dilemma of Global Governance." *The World Bank Economic Review*, 24(2): 171–198.

Kant, Immanuel. 1795. *Perpetual Peace: A Philosophical Essay*. London: S. Sonnenschein.

Kapur, Devesh. 2004. "Remittances: The New Development Mantra?" Issue 29 of G-24 discussion paper series, United Nations Conference on Trade and Development.

Kassem, May. 1999. *In the Guise of Democracy: Governance in Contemporary Egypt*. Reading, UK: Garnet Publishing Limited.

Keefer, Philip. 2007. "Clientelism, Credibility, and the Policy Choices of Young Democracies." *American Journal of Political Science*, 51(4): 804–821.

Keohane, Robert O. 1984. *After Hegemony: Cooperation and Discord in the World Political Economy*. Princeton, NJ: Princeton University Press.

Kepel, Gilles. 2002. *Jihad: The Trail of Political Islam*, trans. Anthony F. Roberts. Cambridge, MA: Harvard University Press.

Kersting, Erasmus, and Christopher Kilby. 2014. "Aid and Democracy Redux." *European Economic Review*, 125–143.

Kurtz, Marcus J., and Sarah Brooks. 2011. "Conditioning the 'Resource Curse': Globalization, Human Capital, and Growth in Oil-Rich Nations." *Comparative Political Studies*, 44(6): 747–770.

Kuziemko, Ilyana, and Eric D. Werker. 2006. "How Much Is a Seat on the Security Council Worth? Foreign Aid and Bribery at the United Nations." *Journal of Political Economy*, 114(5): 905–930.

Laherrere, Jean. 1998. "Development Ration Evolves as True Measure of Exploitation – Evaluating Oil and Reserves." *World Oil*, February: 117–120.

Lai, Brian, and Daniel Morey. "The Impact of Regime Type on the Influence of US Foreign Aid." *Foreign Policy Analysis*, 2(4): 385–404.

Laitin, David. 1999. "Somalia – Civil War and International Intervention." In Barbara F. Walter and Jack Snyder (eds.), *Civil War, Insecurity, and Intervention*. New York: Columbia University Press.

Lake, David. 2009. *Hierarchy in International Relations*. Ithaca, NY: Cornell University Press.

Lancaster, Carol. 2000. *Transforming Foreign Aid: United States Assistance in the 21st Century*. Washington, DC: Peterson Institute.

Langlotz, Sarah, and Niklas Potrafke. 2017. "Does Development Aid Increase Military Expenditure?" Heidelberg University Working Paper 618.

Leblang, David. 2010. "Familiarity Breeds Investment: Diaspora Networks and International Investment." *American Political Science Review*, 104(3): 584–600.

Lemarchand, Rene. 1988. "Beyond the Mad Dog Syndrome." In *The Green and the Black*. Ed. Rene Lemarchand. Bloomington: Indiana University Press.

Levi, Margaret. 1988. *Of Rule and Revenue*. Los Angeles: University of California Press.

Levitt, Steven D., and James M. Synder. 1997. "The Impact of Federal Spending on House Election Outcomes." *Journal of Political Economy*, 105(1): 30–53.

Lewis, I. M. 1988. *A Modern History of Somalia: Nation and State in the Horn of Africa*. Boulder, CO: Westview Press.

Lewis-Beck, Michael S., and Mary Stegmaier. 2000. "Economic Determinants of Electoral Outcomes." *Annual Review of Political Science*, 31: 183–219.

Li, Quan. 2009. "Democracy, Autocracy, and the Expropriation of Foreign Direct Investment." *Comparative Political Studies*, 68(1): 62–74.

Li, Quan, Austin Mitchell, and Eric Owen. 2018. "Why Do Democracies Attract More or Less Foreign Direct Investment? A Meta-Regression Analysis." *International Studies Quarterly*, 62(3): 494–504.

Li, Quan, and Resnick, Adam. 2003. "Reversal of Fortunes: Democracy, Property Rights and foreign direct investment inflows to developing countries." *International Organization*, 57(1): 175–211.

Liang-Fenton, Debra. 2004. *Implementing U.S. Human Rights Policy: Agendas, Policies, and Practices*. Washington, DC: U.S. Institute of Peace.

Liou, Yu-Ming, and Paul Musgrave. 2014. "Refining the Oil Curse: Country-Level Evidence from Exogenous Variations in Resource Income." *Comparative Political Studies*, 47(11): 1584–1610.

Lipset, Seymour O. 1959. "Some Social Requisites of Democracy: Economic Development and Political Legitimacy." *American Political Science Review*, 53(1): 69–105.

Lipsey, Robert E., and Fredrik Sjoholm. 2010. "FDI and Growth in East Asia: Lessons for Indonesia." Institute for Industrial Economics Working Paper 852.

2006. "Foreign Firms and Indonesian Manufacturing Wages: An Analysis with Panel Data." *Economic Development and Cultural Change*, 55(1): 201–221.

Lipsey, Robert E, Fredrik Sjoholm, and Jing Sun. 2010. "Foreign Ownership and Employment Growth in Indonesian Manufacturing." National Bureau of Economic Research Working Paper 15936.

Lipson, Charles. 1985. *Standing Guard: Protecting Foreign Capital in the Nineteenth and Twentieth Centuries*. Berkeley: University of California Press.

Lodewijk, Smets, and Stephen Knack. 2015. "World Bank Policy Lending and the Quality of Public Sector Governance." Policy Research Working Paper Series 7267, World Bank.

Luong, Pauline Jones, and Erika Weinthal. 2010. *Oil Is Not a Curse: Ownership Structure and Institutions in Soviet Successor States*. New York: Cambridge University Press.

Mahdavi, Paasha. 2015. "Explaining the Oil Advantage: Effects of Natural Resource Wealth on Incumbent Reelection in Iran." *World Politics*, 67(2): 226–267.

Mahdavy, Hussein. 1970. "The Patterns and Problems of Economic Development in Rentier States: The Case of Iran." In M. A. Cook (ed.), *Studies in Economic History of the Middle East*. London: Oxford University Press, 428–467.

Malesky, Edmund J. 2008. "Straight Ahead on Red: How Foreign Direct Investment Empowers Subnational Leaders." *Journal of Politics*, 70(1): 97–119.

Malesky, Edmund J., Dimitar D. Gueorguiev, and Nathan M. Jensen. 2015. "Monopoly Money: Foreign Investment and Bribery in Vietnam, a Survey Experiment." *American Journal of Political Science*, 59(2): 419–439.

Marshall, Monty G., Ted Robert Gurr, and Keith Jaggers. 2015. "POLITY IV Project: Political Regime Characteristics and Transitions, 1800–2015 Dataset Users' Manual." Center for Systemic Peace and Societal-Systems Research Inc., University of Maryland.

Marshall, Shana, and Joshua Stacher. 2012. "Egypt's Generals and Transnational Capital." *Middle East Report*, Spring: 12–18.

Melhum, Halvor, Karl Moene, and Ragnar Torvik. 2006. "Institutions and the Resource Curse." *The Economic Journal*, 116(1):1–20.

Meltzer, Allan H., and Scott F. Richards. 1981. "A Rational Theory of the Size of Government." *Journal of Political Economy*, 89(5): 914–927.

Menaldo, Victor. 2016. *The Institutions Curse: Natural Resources, Politics, and Development*. Cambridge, UK: Cambridge University Press.

Meseguer, Covadonga, and Katrina Burgess. 2014. "International Migration and Home Country Politics." *Studies in Comparative International Development*, 49(1):1–12.

Milner, Helen V., and Keiko Kubota. 2005. "Why the Move to Free Trade? Democracy and Trade Policy in the Developing Countries." *International Organization*, 59(4): 107–143.

Milner, Helen V., and Bumba Mukherjee. 2009. "Globalization and Democracy: A Review Essay." *Annual Review of Political Science*, 12: 163–181.

Milner, Helen V., and Dustin H. Tingley. 2010. "The Political Economy of US Foreign Aid: American Legislators and the Domestic Politics of Aid." *Economics and Politics*, 22(2): 200–232.

Moore, Barrington Jr. 1966. *Social Origins of Dictatorship and Democracy: Lord and Peasant in the Making of the Modern World*. Boston: Beacon Press.

Morgenthau, Hans. 1962. "A Political Theory of Foreign Aid." *American Political Science Review*, 56(2): 301–309.

Morrison, Kevin M. 2015. *Nontaxation and Representation: The Fiscal Foundations of Political Stability*. Cambridge: Cambridge University Press.

Mosley, Layna. 2003. *Global Capital and National Governments*. Cambridge: Cambridge University Press.

Mosley, Layna, and David Singer. 2015. "Migration, Labor, and the International Political Economy." *Annual Review of Political Science*, 18: 283–301.

Moyo, Dambisa. 2009. *Dead Aid: Why Aid Is Not Working and How There Is a Better Way for Africa*. New York: Farrar, Straus and Giroux.

Neumayer, Eric. 2003. "What Factors Determine the Allocation of Aid by Arab Countries and Multilateral Agencies?" *Journal of Development Studies*, 39 (4): 134–147.

North, Douglass C. 1991. "Institutions." *Journal of Economic Perspectives*, 5(1): 97–112.

Nunn, Nathan, and Nancy Qian. 2014. "Aiding Conflict: The Impact of U.S. Food Aid on Civil War." *American Economic Review*, 104(6): 1630–1666.

Obama, Barack. 2009. "Memorandum on Promoting Democracy and Human Rights in Cuba." August 13.

O'Donnell, Guillermo A. 1973. *Bureaucratic Authoritarianism*. Berkeley: University of California Press.

O'Donnell, Guillermo. 1979. "Tensions in the Bureaucratic-Authoritarian State and the Question of Democracy." In David Collier (ed.), *The New Authoritarianism in Latin America*. Princeton, NJ: Princeton University Press, 285–318.

O'Mahony, Angela. 2013. "Political Investment: Remittances and Elections." *British Journal of Political Science*, 43(4): 799–820.

Organisation for Economic Co-operation and Development (OECD). 2018. "Foreign Direct Investment Statistics: Data, Analysis and Forecast." www.oecd.org/corporate/mne/statistics.htm

Pandya, Sonal S. 2010. "Labor Markets and Demand for Foreign Direct Investment." *International Organization*, 64(3): 389–409.

2016. "Political Economy of Foreign Direct Investment: Globalized Production in the Twenty-First Century." *Annual Review of Political Science*, 19: 455–475.

Pepinsky, Thomas. 2008. "Capital Mobility and Coalitional Politics: Authoritarian Regimes and Economic Adjustment in Southeast Asia." *World Politics*, 60(3): 438–474.

Perez-Armendariz, Clarisa, and David Crow. 2010. "Do Migrants Remit Democracy? International Migration, Political Beliefs and Behavior in Mexico." *Comparative Political Studies*, 43(1): 119–148.

Peters, Anne Mariel, and Peter W. Moore. 2009. "Beyond Boom and Bust: External Rents, Durable Authoritarianism, and Institutional Adaption in the Hashemite Kingdom of Jordan." *Studies in Comparative International Development*, 44: 256–285.

Pevehouse, Jon C. 2005. *Democracy from Above? Regional Organizations and Democratization*. New York: Cambridge University Press.

Pfutze, Tobias. 2012. "Does Migration Promote Democratization? Evidence from the Mexican Transition." *Journal of Comparative Economics*, 40(2): 159–175.

Pinto, Pablo. 2013. *Partisan Investment in the Global Economy*. Cambridge: Cambridge University Press.

Pinto, Pablo, and Boliang Zhu. 2016. "Fortune or Evil? The Effect of Inward Foreign Direct Investment on Corruption." *International Studies Quarterly*, 60(4): 693–705.

Poulsen, Lauge N. Skovgaard. 2015. *Bounded Rationality and Economic Diplomacy*. New York: Cambridge University Press.

Profeta, Paola, and Simona Scabrosetti. 2010. *The Political Economy of Taxation: Lessons from Developing Countries*. Northampton, MA: Edward Elgar Publishing.

Przeworski, Adam, Michael E. Alvarez, Jose Antonio Cheibub, and Fernando Limongi. 2000. *Democracy and Development; Political Institutions and Well-Being in the World, 1950–1990*. New York: Cambridge University Press.

Ramsay, Kristopher. 2011. "Revisiting the Resource Curse: Natural Disasters, the Price of Oil, and Democracy." *International Organization*, 65: 507–529.

Ratha, Dilip. 2007. "Leveraging Remittance for Development." Working Paper, Development Prospects Group, World Bank.

Rawson, David. 1994. "Dealing with Disintegration: U.S. Assistance and the Somali State." In Ahmed I. Samatar (ed.), *The Somali Challenge: From Catastrophe to Renewal?* Boulder, CO: Lynne Reinner, 147–187.

Robinson, James A., Ragnar Torvik, and Thierry Verdier. 2006. "Political Foundations of the Resource Curse." *Journal of Development Economics*, 79(2): 447–468.

Robison, Richard. 1986. *Indonesia: The Rise of Capital*. Sydney: Allen and Unwin.

1988. "Authoritarian States, Capital-Owning Classes, and the Politics of New Industrializing Countries: The Case of Indonesia." *World Politics*, 41(1): 52–74.

Rodden, Jonathan. 2004. "Comparative Federalism and Decentralization: On Meaning and Measurement." *Comparative Politics*, 36(4): 481–500.

Rodrik, Dani. 1995. "Political Economy of Trade Policy." In Gene M. Grossman and Kenneth Rogoff (eds.), *Handbook of International Economics*, Vol. 3. Amsterdam: North-Holland, 1457–1494.

1998. "Why Do More Open Economies Have Bigger Governments?" *Journal of Political Economy*, 106(5): 997–1032.

2011. *The Globalization Paradox: Democracy and the Future of the World Economy*. New York: W. W. Norton and Company.

Roodman, David. 2007. "The Anarchy of Numbers: Aid, Development, and Cross-Country Empirics." *World Bank Economic Review*, 21(2): 255–277.

Rose-Ackerman, Susan, and Jennifer Tobin. 2011. "When BITs Have Some Bite: The Political-Economic Environment for Bilateral Investment Treaties." *The Review of International Organizations*, 6(1): 1–32.

Rosendorff, B. Peter, and Kongjoo Shin. 2011. "Importing Transparency: The Political Economy of BITs and FDI Flows." New York University Working Paper.

Ross, Michael L. 2001. "Does Oil Hinder Democracy?" *World Politics*, 53(3): 325–361.

2015. "What Have We Learned About the Resource Curse?" *Annual Review of Political Science*, 18: 239–259.

2012. *The Oil Curse: How Petroleum Wealth Shapes the Development of Nations*. Princeton, NJ: Princeton University Press.

2013. "The Politics of the Resource Curse: A Review." Working Paper.

Roubini, Nouriel, and Jeffrey Sachs. 1989. "Political and Economic Determinants of Budget Deficits in the Industrial Economies." *European Economic Review*, 33: 903–938.

Rudra, Nita. 2005. "Globalization and the Strengthening of Democracy in the Developing World." *American Journal of Political Science*, 49(4): 704–730.

Rudra, Nita, Meir Alkon, and Siddarth Joshi. 2018. "FDI, Poverty, and Politics of Potable Water Access." *Economics and Politics*, 30(3): 366–393.

Sarkees, Meredith Reid, and Frank Wayman. 2010. *Resort to War: 1816–2007*. Washington, DC: CQ Press.

Schumpeter, Joseph. 1954 (1918). "The Crisis of the Tax State." In (A. Peacock, ed.), *International Economic Papers: Translations Prepared for the International Economic Association*. New York: Macmillan, 5–38.

Scott, James M., and Carie A. Steele. 2011. "Sponsoring Democracy: The United States and Democracy Aid to the Developing World, 1988–2001." *International Studies Quarterly*, 55: 47–69.

Sen, Amartya. 1999. *Development as Freedom*. New York: Oxford University Press.

Shehata, Dina. 2011. "The Fall of the Pharaoh: How Hosni Mubarak's Reign Came to an End." *Foreign Affairs*, 90(3): 26–32.

Shi, Weiyi. 2014. *The Political Economy of China's Outward Direct Investments*, PhD dissertation, Department of Political Science, University of California San Diego.

Simpser, Alberto, Lauren Duquette-Rury, Jose A. H. Company, and Juan F. Ibarra. 2016. "The Political Economy of Social Spending by Local Government: A Study of the 3x1 Program in Mexico." *Latin American Research Review*, 51(1): 62–83.

Singer, David. 2012. "The Family Channel: Migrant Remittances and Government Finance." MIT Political Science Research Paper 2012-23.

SIPRI Military Expenditure Database. http://milexdata.sipri.org/

Smith, Alastair. 2008. "The Perils of Unearned Income." *Journal of Politics*, 70 (3): 780–793.

Soesastro, Hadi, and M. Chatib Basri. 1998. "Survey of Recent Developments." *Bulletin of Indonesian Economic Studies*, 34 (April): 3–54.

Sovey, Allison J., and Donald P. Green. 2011. "Instrumental Variables Estimation in Political Science: A Reader's Guide." *American Journal of Political Science*, 55(1): 188–200.

Staats, Joseph L., and Glen Biglaiser. 2012. "Foreign Direct Investment in Latin America: The Importance of Judicial Strength and Rule of Law." *International Studies Quarterly*, 56(1): 193–202.

Stock, James H., and Francesco Trebbi. 2003. "Who Invested Instrumental Variable Regression?" *Journal of Economic Perspectives*, 17(3): 177–194.

Stock, James H., Jonathan H. Wright, and Motohiro Yogo. 2002. "A Survey of Weak Instruments and Weak Identification in Generalized Method of Moments." *Journal of Business and Economic Statistics*, 20(4):518–529.

Svolik, Milan W. 2012. *The Politics of Authoritarian Rule*. New York: Cambridge University Press.

Takii, Sadayuki, and Eric D. Ramsetter. 2005. "Multinational Presence and Labor Productivity Differentials in Indonesian Manufacturing, 1975–2001." *Bulletin of Indonesian Studies*, 41: 221–242.

Temenggung, Della. 2006. "Productivity Spillovers from Foreign Direct Investment: Indonesian Manufacturing Industry's Experience, 1975–2001." Working Paper, Australia National University, Department of Economics. http://degit.sam.sdu.dk/papers/degit_12/C012_048.pdf

Therein, Jean-Pierre, and Alain Noel. 2000. "Political Parties and Foreign Aid." *American Political Science Review*, 94(1): 151–162.

Thomas, Kenneth D., and Y. Panglaykim. 1973. *Indonesia: The Effect of Past Policies and President Suharto's Plans for the Future*. Melbourne: Committee for Economic Development of Australia.

Tilly, Charles. 1992. *Coercion, Capital, and European States, AD 990–1992*. Cambridge, MA: Blackwell.

Tomz, Mike. 2007. *Reputation and International Cooperation: Sovereign Debt across Three Centuries*. Princeton, NJ: Princeton University Press.

Tordo, Silvana, Brandon S. Tracy, and Noora Arfaa. 2011. "National Oil Companies and Value Creation." World Bank Working Paper 218. Tornell, Aaron, and Philip R Lane. 1999. "The Voracity Effect." *American Economic Review*, 89(1): 22–46.

Treisman, Daniel. 2000. "The Causes of Corruption: A Cross-National Study." *Journal of Public Economics*, 76(3): 399–457.

  2007. "What Have We Learned about the Causes of Corruption from Ten Years of Cross-National Empirical Research?" *Annual Review of Political Science*, 10(1): 211–244.

Tufte, Edward R. 1975. "Determinants of the Outcomes of Midterm Congressional Elections." *American Political Science Review*, 69(3): 812–826.

United Nations. 2006. "World Investment Report 2006: FDI from Developing Developing and Transition Economies: Implications for Development." New York and Geneva: United Nations Conference on Trade and Development.

Vreeland, James R. 2008. "The Effect of Political Regime on Civil War: Unpacking Anocracy." *Journal of Conflict Resolution*, 52(3): 401–425.

Ward, Ken. 1973. "Indonesia's Modernization: Ideology and Practice." In Rex Mortimer (ed.), *Showcase State: The Illusion of Indonesia's Accelerated Modernization*. Sydney: Angus and Robertson.

Werker, Eric D. 2012. "The Political Economy of Bilateral Foreign Aid." In Gerrard Caprio (ed.), *Handbook of Safeguarding Global Financial Stability: Political, Social, Cultural, and Economic Theories and Models*. London: Academic Press, 47–58.

Werker, Eric, and Faisal Z. Ahmed. 2008. "What Do Non-governmental Organizations Do?" *Journal of Economic Perspectives*, 22(2): 73–92.

Wintrobe, Ronald. 1998. *The Political Economy of Dictatorship*. New York: Cambridge University Press.

Wood, Reed M., and Mark Gibney. 2010. "The Political Terror Scale (PTS): A Re-introduction and a Comparison to CIRI." *Human Rights Quarterly* 32(2): 367–400.

World Bank. 2010. *World Development Indicators, 2010*. Washington, DC: World Bank.

Wright, Joseph. 2009. "How Can Foreign Aid Foster Democratization in Authoritarian Regimes?" *American Journal of Political Science*, 53(3): 552–571.

Zhu, Boliang. 2017. "MNCs, Rents, and Corruption: Evidence from China." *American Journal of Political Science*, 61(1): 84–99.

Index

Africa. *See also specific countries*
  foreign aid from China, 1–2
Amin, Idi, 131
Arab oil embargo, 103
Arab Spring. *See also specific countries*
  remittances as influence on, 122–125
    executive constraints from, 123–125
    patronage politics and, 122–123
Asian Financial Crisis, 149
ATE. *See* average treatment effect
Augustus (Emperor), 54
authoritarian regimes. *See also* autocracies
  cross-sectional variations in, 103
  data analysis of, 110–112
    executive constraints in, 111
    independent variables in, 111–112
    measurement indices, 110–111
  oil discoveries in, 130
    authoritarian collapse and, 149–150
  political survival of, 111
  remittances to, 103–104
    identification strategies for, 107–110
  temporal variations in, 103
autocracies. *See also specific countries*
  under Barre, 82–84
  closed-off, 157
  cofounders' role in, 36–40
  through cross-border flows, 40–41
  FDI in, 2–3
  foreign rents in, 64
  hybrid, 156
  income effects in, 74–75
  international capital in, 15–16, 30–40
    theoretical model for, 69–75

  leader exit in, 38, 39
    fixed effects of, 38–40
  leader tenure in, 33–36
    in strong autocracies, 33, 34
    in weak autocracies, 34
  military spending and, 58
  New Order regime, in Indonesia, 62
  per capita transfers in, 64
  performance legitimacy and, 53–55
  political repression and, 50, 51
  political survival of, 32–40
  raw data patterns in, 32–36
  remittances to, 105, 112–115, 125
    endogeneity bias in, 106–107
    identification strategies for, 107–110
    migration patterns and, 106–107
    political engagement through, 105
    causal validity through, 112–113
    strengthening of political power through, 112–115
  rentier effects in, 74–75
  strong, 32–33
    leader tenure in, 33, 34
  substitution effects in, 74–75
  theoretical models for
    backward induction solutions, 73–74
    construction of, 69–70
    for international capital, 69–75
    interpretation of, 74–75
    players and preferences in, 70–72
    sequence of moves in, 72
  in Triple Alliance, 62
  U. S. foreign aid to, 98–102
    political survival of, 100–102

autocracies (cont.)
  weak, 34
  welfare goods in, 56
average treatment effect (ATE), of remittances, 21

Bahrain, Arab Spring in, 122–125
Bangladesh, as democracy, 35
Barre, Siad, 1–2, 81–86, 102. *See also* Somalia
  autocratic rule under, 82–84
  scientific socialism and, 81–83
bias. *See* endogeneity bias
bilateral aid, from U. S., 79, 80, 88
  determinants of, 93–95
  probability of, 89
  in Somalia, 84
bilateral donors, of foreign aid, 25
bilateral investment treaties (BITs), 163–164
borrowing. *See* sovereign borrowing and debt
Brazil
  FDI in, 2–3
  MNCs in, 12
  Triple Alliance in, 62, 130

capital. *See* international capital
Castro, Fidel, 2
Castro, Raul, 2
China, foreign aid to Africa, 1–2
Ciputra, 148
civil conflicts, 145
  riots, 121
clientelism, 105
closed-off autocracies, 157
Cold War, U. S. foreign aid during, 81, 84–85
colonialism, FDI influenced by, 12
conditional effects, of revenue generation, 48–49
conditional theory, of international capital, 7
Congress, U. S., foreign aid through, 86–88
corruption, political
  definition of, 59
  FDI as influence on, 13
  ICRG measurement of, 59
  international capital and, 157
  MNCs and, 157
  in public spending, 57–59
    in democracies, 60
  through patronage, 59
  remittances as funding source for, 117–119
    ICRG corruption index measures, 118
    patronage-based, 118
coups, political, 145

Cuba
  during Castro regimes, 2
  foreign remittances to, 2

debt. *See* sovereign borrowing and debt
democracies
  attitudes toward foreign aid, 160
  Bangladesh as, 35
  globalization of, 163
  institutional scores of, 30–31
  military spending in, 58
    oil discoveries as influence on, 135
  MNCs in, 11, 12
  performance legitimacy in, 53–55
  political corruption in, 60
  political repression in, 50, 51
  politics in, 159–160
  public sector compensation in, 54
  trade liberalization in, 15, 162
  welfare goods in, 56, 57
democratic peace theory, 14–15
developing countries, international capital in, 23
development. *See* international development
Development with Freedom (Human Rights Watch), 1
dictatorships
  FDI under, 2–3
  foreign rents in, 64
  international capital under, 7
  leader exit in, 44
  oil discoveries in, 150–151
  political repression in, 50
  tinpot, 55
donors, of foreign aid
  bilateral, 25
  characteristics of, governance effects influenced by, 165
  conditionalities of, 78
  policy concessions for, 78
  political benefits for, 25
  political intent of, 9, 19
  to Somalia, 84. *See also* Somalia

economic growth, foreign aid as influence on, 78
  through U. S. foreign aid, 92, 102
Egypt
  Arab Spring in, remittances as influence on, 122–125
    executive constraints from, 123–125
    patronage politics and, 122–123

Index

MNCs in, 12
U. S. foreign aid to, 81
endogeneity bias, 86, 106–107
endogenous effects
   foreign remittances and, 20
   of international capital, 66, 164
   on political survival, 76
Ethiopia, foreign aid to
   as political weapon within Ethiopia, 1
   from U. S., 1
executive constraints index, 98–99
executive powers, political repression through, 55
expenditure-switching mechanisms. *See* substitution effects

FDI. *See* foreign direct investment
federal systems, revenue generation in, 46
foreign aid. *See also* bilateral aid; donors; U. S. foreign aid
   definition of, 2, 25–26
   to democracies, public attitudes toward, 160
   economic growth influenced by, 78
   to Ethiopia
      as political weapon within Ethiopia, 1
      from U. S., 1
   through FDI, 2–6
      in Brazil, 2–3
      cumulative flows of, from 2000–2015, 5
      government structure as factor in, 2–3
      international cumulative flows of, from 2000–2015, 4–6
      investment gaps and, 3
   through foreign remittances, 4
      to Cuba, 2
      to Philippines, 6
   global calculations of, 27, 28
      top recipients in, 29
   IBRD and, 78
   IDA and, 78
   misuse of, 2
   NGOs and, 79
   as ODA, 25–26
   policy concessions for, 78
   politics and, 77–86
   regime type influenced by, 12–13
   resource curse and, 16
   to Somalia, 1–2
   varieties of, 77–79
foreign direct investment (FDI). *See also* oil discoveries
   BITs and, 163–164

under colonial rule, 12
composition of, 6
definition of, 27
efficiency-seeking, 128
foreign aid through, 2–6
   in Brazil, 2–3
   cumulative flows of, from 2000–2015, 5
   government structure as factor in, 2–3
   international cumulative flows of, from 2000–2015, 4–6
   investment gaps and, 3
global calculations of, 28
   top recipients in, 29
green field investments, 130
indirect government effects of, 27
market-seeking, 128
MNCs and, 11–12, 128, 129
political corruption influenced by, 13
politics of, 127–130
   government incentives in, 128
   motivations in, 128
public finances through, 62
remittance flows influenced by, 13
rentier effect in, 7, 42, 68
rents and, 129–130
research on, 11–13
resource-seeking, 128
as sequential process, 42
types of, 127–129
foreign rents, in nondemocracies, 64
Freedom House, 46

globalization
   of democracy, 163
   institutionalization of, 162–163
   international capital influenced by, 14–15, 160–164
Goldscheid, Rudolf, 44
governments. *See also* public finances and spending; regime types; revenues; *specific types of government*
   private consumption by, 55–59
   patronage as, 121
   public consumption by, 55–59

Hasan, Bob, 148
Hawala method, of foreign remittances, 26
Human Rights Watch, 1
hybrid autocracies, 156

IBRD. *See* International Bank for Reconstruction and Development

ICRG corruption index. *See* International Country Risk Guide corruption index
IDA. *See* International Development Association
income effects, 7, 42, 68
  in autocracies, 74–75
  of nontax income, 61
Indonesia. *See also* New Order regime
  FDI in, 2–3
  MNCs in, 12
institution curse, 16
International Bank for Reconstruction and Development (IBRD), 78
international capital
  benefits of, 164–166
  causal effects of, 17–18
  composition of, 30
  conditional theory of, 7
  corruption and, 157
  countervailing effects of, 65–66
  country averages of, 25
  definition of, 25–27
  in democratic peace theory, 14–15
  under dictatorships, 7
  empirical analysis of, 65–67, 152–156
    causal identification in, 66–67
    countervailing effects in, 65–66
    endogenous variation in, 66
    evaluating channels in, 66–67
    evidence in, 153–155
    exogenous variation in, 66
    inferences in, 155–156
  endogenous effects of, 66, 164
  globalization as influence on, 14–15, 160–164
  institution curse and, 16
  institutional change influenced by, 9
  international development influenced by, 16–17
  measurement of, 25–32
  methodological approach to, 6–8
    dependent variables in, 6
    income effects in, 7, 42, 61, 68
    rational political actors in, 6
    rentier effect in, 7, 42, 68
    substitution effect in, 7, 42
  microfoundations and, 15
  as nontax income
    channels for, 61–64
    income effects, 61
    in non-democracies, 63–64
    through rent creation, 62
  substitution effects, 61–62
  for political survival, 64
  regime type influenced by, 15–16. *See also* autocracies
  in contemporary politics, 16–17
  research on, 8–14
    FDI and, 11–13
    foreign aid in, role of, 8–9
    regression processes, 18
    remittances and, 9–11
    sample of developing countries in, 23
    sovereign borrowing and debt in, 13–14
  resource curse and, 16
  revenue-based theory of, 67–68
  subnational analysis of, 156–159
  types of, 25
  variation in, 27–30
    spatial, 28–29
    temporal, 27–28
  welfare implications of, 75, 166
International Country Risk Guide (ICRG) corruption index, 59, 118
international development, international capital as influence on, 16–17
International Development Association (IDA), 78

Jordan, Arab Spring in
  remittances as influence on, 122–125
  executive constraints from, 123–125
  patronage politics and, 122–123

Kant, Immanuel, 14
Kennedy, John F., 131

leader exit
  in autocracies
    fixed effects in, 38–40
    international capital as influence on, 38, 39
  in dictatorships, 44
  foreign interventions in, 49
  oil discoveries and, 145–146
leader tenure, in autocracies, international capital as influence on, 33–36
  in strong autocracies, 33, 34
  in weak autocracies, 34
liberalism. *See also* trade liberalization
  through U. S. foreign aid disbursement, 80
Liem Sioe Liong, 148
loyalty, accumulation of, 52–55

Index

Mexico, remittances to, 158
microfoundations, 15
militarized interstate dispute (MID), 138
military spending, 56–57
  in autocracies, 58
  in democracies, 58
    oil discoveries as influence on, 135
  FDI for, 126–127
  oil discoveries as influence on, through FDI, 133–138, 141
    in autocracies, 137, 141
    composition of military spending, 137–138
    in democracies, 135
    in nondemocracies, 136
MNCs. *See* multinational corporations
modernization theory, 121, 149
multinational corporations (MNCs)
  in Brazil, 12
  corruption and, 157
  in democracies, 11, 12
  in Egypt, 12
  FDI and, 11–12, 128, 129
  in Indonesia, 12
Muslim countries. *See also specific countries*
  remittances to, 107, 109
    determinants of, 112–113, 114
    causal validity through, 112–113
    pre-treatment characteristics in, 112, 113

New Order regime, in Indonesia, 62, 130
  FDI during, 146–150
    authoritarian collapse in, 149–150
    capital outflows in, 149–150
    as foreign capital, 146–149
    import substitution industrialization in, 147
    through military takeover, 147
    in modernization theory, 149
nongovernmental organizations (NGOs), 79
non-Muslim countries, remittances to
  determinants of, 112–113, 114
    causal validity through, 112–113
    pre-treatment characteristics in, 112, 113
nontax income, international capital as
  channels for, 61–64
  income effects, 61
  in non-democracies, 63–64
  through rent creation, 62
  substitution effects, 61–62
official development assistance (ODA), 25–26

oil discoveries, FDI in, 130–145
  in authoritarian regimes, 130
    collapse of, 149–150
  data analysis of, 132–133
  in dictatorships, 150–151
  evaluation of channels, 138–145
    FDI surges after, 138–141
  identification strategies, 130–132, 135–137
  military spending influenced by, 133–138, 141
    in autocracies, 137, 141
    composition of, 137–138
    in democracies, 135
    in nondemocracies, 136
  natural disasters and, 135
  under New Order regime, in Indonesia, 146–150
    authoritarian collapse in, 149–150
    capital outflows in, 149–150
    foreign capital in, attraction of, 146–149
  political destabilization from, 144
  political survival and, 145–146
    through civil conflicts, 145
    of leaders, 145–146
    political coups and, 145
    rent allocation and, 145
  security competition after, 139
  success/failures in oil exploration and, 133–135, 137

patronage, 53
  public finances and
    political corruption and, 59
    private government consumption, 121
  remittances and, 118
patronage politics, during Arab Spring, 122–123
per capita transfers, in autocracies, 64
performance legitimacy, of regime types, 53–55
Philippines, foreign remittances to, 6
political coups. *See* coups
political engagement, remittances as influence on, 105
  in autocracies, 105
  clientelism and, 105
political freedom, Freedom House measures of, 46
political institutions, leader tenure and, 32
political rights, 47
  U. S. foreign aid as influence on, 93–96

political survival. *See also* leader exit; leader tenure; revenues
  of authoritarian regimes, 111
  of autocracies, 32–40
  endogenous effects on, 76
  foundations of, 43–49
    in dictatorships, 43–44
  international capital for, 64
  methodological approach to, 76–77
  oil discoveries and, 145–146
    civil conflicts and, 145
    of leaders, 145–146
    political coups and, 145
    rent allocation and, 145
  by regime type, 32
  through remittances, 115
  in selectorate theory, 117
  resource income and, 48–49
  strategies for, 49–55
    through accumulation of loyalty, 52–55
    through patronage, 53
    through performance legitimacy, 53–55
    through political repression of citizens, 49–52
    in selectorate theory, 52–53
    U. S. foreign aid as influence on, 101
  in autocracies, 100–102
political terror, 51
politics, as concept, 24
POLITY index, 99
private government consumption, 55–59
  patronage as, 121
productivity spillovers, 149
public finances and spending, 55–60. *See also* revenues; substitution effects
  corruption and, 57–59
    in democracies, 60
    through patronage, 59
  through FDI, 62
  military spending, 56–57
    in autocracies, 58
    in democracies, 58
  political survival and, 60
  private government consumption and, 55–59
    patronage as, 121
  public government consumption and, 55–59
  through remittances, 105–106, 117
  in selectorate theory, 55–56
  welfare goods, 56
    in autocracies, 20, 56
    in democracies, 56, 57

public government consumption, 55–59
public sector compensation, in democracies, 54

Rahardja, Hendra, 148
rational political actors, 6
regime types. *See also* autocracies; democracies; dictatorships
  international capital as influence on, 15–16
  in contemporary politics, 16–17
  performance legitimacy of, 53–55
  political survival measure by, 32
  tenure of leaders and, 32
remittances, 4, 104–106
  Arab Spring influenced by, 122–125
    executive constraints and, 123–125
    patronage politics and, 122–123
  ATE of, 21
  to authoritarian regimes, 103–104
    identification strategies for, 107–110
  to autocracies, 112–115, 125
    endogeneity bias in, 106–107
    identification strategies for, 107–110
    migration patterns and, 106–107
    political engagement in, 105
    causal validity through, 112–113
    strengthening of political power through, 112–115
  corruption funded by, 117–119
    ICRG corruption index measures, 118
    patronage-based, 118
  to Cuba, 2
  default risks influenced by, 11
  definition of, 26, 165
  endogeneity effects from, 20
  evaluation of channels of, 116–122
    in modernization theory, 121
    sovereign borrowing mechanisms, 119–121
    through substitution effects, 116
  FDI influenced by, 13
  global calculations of, 28
    top recipients in, 29
  Hawala method, 26
  identification strategies for, 107
    in authoritarian regimes, 107–110
    in autocracies, 107–110
    through causal inferences, 110
    exclusion restrictions and, 110
    instrumental variables in, 108–110
  in international capital research, 9–11
  to Mexico, 158
  to Muslim countries, 107, 109

determinants of, 112–113, 114
causal validity through, 112–113
pre-treatment characteristics in, 112, 113
to non-Muslim countries
determinants of, 112–113, 114
causal validity through, 112–113
pre-treatment characteristics in, 112, 113
to Philippines, 6
political engagement and, 105
in autocracies, 105
clientelism and, 105
political survival through, 115
in selectorate theory, 117
public finances and, 105–106, 117
quasi-natural experiments for, 106–112
riots influenced by, 121
as unearned foreign income, 10
World Bank data on, 26–27
rent allocation, 145
rent creation, nontax income through, 62
rentier effects, 7, 42, 68
in autocracies, 74–75
rentier state, 47
repression, political
in autocracies, 50, 51
classifications of, 51
definitions of, 51
in democracies, 50, 51
in dictatorships, 50
through executive powers, 55
measurement scales for, 50–51
political survival through, 49–52
revenue generation influenced by, 49
U. S. foreign aid as factor in, 93–96
evaluation of channels for, 96–98
loss of political rights and, 94, 93–96
in Somalia, 81–86
resource curse, 16
resource income, political survival and, 48–49
revenues, generation of
political repression as influence on, 49
political survival influenced by
conditional effects of, 48–49
in federal systems, 46
through nontax sources, 45–48
political rights influenced by, 47
through political use of revenue, 44–45
primacy of, 44–46
through taxation, 44, 47
for war, 46
rights. *See* political rights
riots, remittances and, 121

selectorate theory, 52, 117
exclusion risks in, 52–53
public finances and spending in, 55–56
Senate, U. S., foreign aid through, 88–89
Soerjadjaja, Willem, 148
Somalia
in Arab League, 85
as autocracy, 82–84
clannism in, 82
foreign aid to, 1–2
nationalism in, 82
scientific socialism in, 81–83
U. S. foreign aid to, 81–86, 102
Barre and, 81–86
as bilateral aid, 84
donors of, 84
endogeneity bias in, 86
political repression as result of, 81–86
sovereign borrowing and debt, 13–14
remittances influenced by, 119–121
strong autocracies, 32–33
leader tenure in, 33, 34
substitution effects, 7, 42, 68
in autocracies, 74–75
of nontax income, 61–62
remittances and, 116
Suharto, 146–150. *See also* New Order regime
during Asian Financial Crisis, 149
political demise of, 149–150

Tan Siong Ke, 148
taxation
rent creation and, 62
revenue generation through, 44, 47
tinpot states, 55
trade liberalization, in democracies, 15, 162
Triple Alliance, 62, 130
Tunisia, Arab Spring in, 122–125

U. S. *See* United States
U. S. foreign aid, 79–102
to autocracies, 98–102
political survival of, 100–102
as bilateral aid, 79, 80, 88
determinants of, 93–95
probability of, 89
in Somalia, 84
during Cold War, 81, 84–85
disbursement of, variations in, 90–93
data analysis for, 93
endogenous variables for, 91–91
exogenous variables for, 91–91

U. S. foreign aid (cont.)
　instrumental variables for, 90–93
　studies on, 91
　for economic growth, 92, 102
　to Egypt, 81
　to Ethiopia, 1
　evaluation of channels for, 96–98
　　for political repression, 96–98
　　reduction of tax effort, 96–98
　executive constraints index for, 98–99
　exogeneity factors, 88–89
　frequency of, by recipient, 89, 90
　identification strategies for, 86–93
　　legislative determinants in, 86–90
　legislative fragmentation for, 86–88, 155
　　in Congress, 86–88
　　in Senate, 88–89
　political liberalization through, 80
　political repression as result of, 93–96
　　evaluation of channels for, 96–98
　　loss of political rights, 93–96
　　in Somalia, 81–86
　political survival through, 101
　　in autocracies, 100–102
　politics of, 79–81
　　geostrategic considerations in, 81
　POLITY index for, 99
　to Somalia, 81–86, 102
　　Barre and, 81–86
　　as bilateral aid, 84
　　donors of, 84
　　endogeneity bias in, 86
　　political repression as result of, 81–86
　volatility of, 79
United States (U. S.)
　congressional considerations for foreign aid, 86–88
　Senate considerations for foreign aid, 88–89

war, revenue generation for, 46
weak autocracies, 34
welfare goods, 56
　in autocracies, 56
　in democracies, 56, 57
　international capital and, 75, 166
World Bank, on foreign remittances, 26–27

Yemen, Arab Spring in, 122–125
Yom Kippur War, 103

Zenawis, Meles, 1